Science Networks. Historical Studies

Birkhäuser

Science Networks. Historical Studies
Founded by Erwin Hiebert and Hans Wußing
Volume 51

Edited by Eberhard Knobloch, Helge Kragh and Volker Remmert

More information about this series at
http://www.birkhauser-science.com/series/4883

Vincenzo De Risi

Leibniz on the Parallel Postulate and the Foundations of Geometry

The Unpublished Manuscripts

Birkhäuser

Vincenzo De Risi
MPI for the History of Science
Berlin, Germany

ISSN 1421-6329 ISSN 2296-6080 (electronic)
Science Networks. Historical Studies
ISBN 978-3-319-19862-0 ISBN 978-3-319-19863-7 (eBook)
DOI 10.1007/978-3-319-19863-7

Library of Congress Control Number: 2015960187

Mathematics Subject Classification 2010: 01-02, 01A45, 01A50, 02-03, 51-03

Springer Cham Heidelberg New York Dordrecht London

Cover illustration: From Waller Ms de-00215, August Beer: Über die Correction des Cosinusgesetzes bei der Anwendung des Nicol'schen Prismas in der Photometrie, after 1850. With friendly permission by The Waller Manuscript Collection (part of the Uppsala University Library Collections).

Printed on acid-free paper

Springer International Publishing AG Switzerland is part of Springer Science+BusinessMedia (www.birkhauser-science.com)

Foreword

In this book, I provide an edition of Leibniz' writings on the theory of parallels, which include several attempts to prove the Parallel Postulate. A few of these papers were published by Leibniz himself in his lifetime, while others were only printed in modern editions of his mathematical works. The most important essays, however, were still unpublished, and I have transcribed them from the Leibnizian manuscripts in the *Niedersächsische Landesbibliothek* in Hannover. Given the enormous amount of Leibnizian papers preserved in the *Leibniz-Archiv*, I cannot claim to have found all the relevant material, and to have a complete picture of Leibniz' endeavors in this direction we have to wait for the publication of the related volumes by the Academy Edition of Leibniz' *Werke*; since however the scholars at the *Leibniz-Archiv* have just begun the first surveys of Leibniz' geometrical writings after 1676, it is likely that a full edition will require many years. In any case, I am confident that the present collection of papers on the theory of parallels is comprehensive enough to give a quite good idea of Leibniz' work in this field.

Most of the texts presented here are sections and paragraphs of longer essays on the foundations of geometry, while a few others are self-contained notes and remarks on the Parallel Postulate that Leibniz penned from time to time. Given the highly fragmentary character of these drafts and private notes, their meaning and significance may easily be missed and reading them requires a careful study of Leibniz' intellectual development and environment. To this end, I introduce them with an essay commenting the most relevant passages and outcomes, while dealing with the history of the attempts to prove the Parallel Postulate at the time of Leibniz, the main epistemological tenets of Leibniz' philosophy of geometry, and the historical reception of Leibniz' ideas on the subject. On the one hand, my introductory essay is strictly related to my previous book *Geometry and Monadology*, published in this Birkhäuser series in 2007; and while the latter book dealt with Leibniz' philosophy of space and *metaphysical* foundations of geometry, the present essay complements those researches expounding Leibniz' geometrical *epistemology* (albeit from a very specific perspective). On the other hand, this volume may also be read in connection with my commented editions of Saccheri and Lambert on the theory of parallel lines (both published by Birkhäuser), and the three books together offer a comprehensive account of the prehistory of non-Euclidean geometry in the eighteenth century.

I would like to thank the *Leibniz-Archiv* and the *Niedersächsische Landesbibliothek* for allowing me to read, transcribe and publish Leibniz' manuscripts on the theory of parallels. My deepest gratitude goes to Siegmund Probst, whose help in finding and deciphering Leibniz' papers was invaluable for the present edition.

I began to work on Leibniz' theory of parallels in 2009, while I was Alexander von Humboldt Fellow at the Technische Universität Berlin. I would like to thank the Alexander von Humboldt Stiftung for financial support, and my generous host in Berlin, Eberhard Knobloch, who also carefully read and commented on the first draft of this book. His suggestions and advice saved me from several mistakes and considerably enhanced the final version. I am also very grateful to Richard Arthur, Gideon Freudenthal, Mattia Mantovani, and Victor Pambuccian, whose illuminating remarks on further drafts of the volume were crucial to my understanding of several passages.

My studies were presented and discussed in a few seminars from 2010 onwards, in Paris, Hannover, Pisa, Urbino, Leipzig, Ghent, and Princeton, and I am grateful to all the participants who helped me in understanding Leibniz' mathematics and epistemology; in particular, I mention here Herbert Breger, Daniel Garber, Tal Glezer, Pierluigi Graziani, Jürgen Jost, Massimo Mugnai, Enrico Pasini, Francesco Piro, and David Rabouin, whose comments and remarks substantially improved the present study.

Finally, I would like to thank Fred Sengmueller and James Garahan for a linguistic revision of the manuscript, David Merry for having helped me with the translation of Leibniz' texts, and Chiara Fabbrizi for the general editing.

This book is dedicated to my mother Laura, whose unfailing care made everything possible.

Berlin, January 2015

Contents

1 **Introduction** . . . 3

2 **The Theory of Parallel Lines in the Age of Leibniz** . . . 7
- 2.1 From Antiquity to the Renaissance. . . . 7
- 2.2 Jesuit and French Attempts in the Seventeenth Century. . . . 11
- 2.3 Italian and British Attempts in the Seventeenth Century. . . . 14

3 **Leibniz' Epistemology of Geometry and the Parallel Postulate** . . . 21
- 3.1 Proving Axioms. . . . 21
- 3.2 Leibniz' Theory of Definition. . . . 31
- 3.3 Geometry as the Science of Space. . . . 40
- 3.4 Philosophy and the Parallel Postulate. . . . 49

4 **Leibniz' Attempts to Prove the Parallel Postulate** . . . 57
- 4.1 Preliminary Definitions. . . . 57
- 4.2 Leibniz' Youth and the Parisian Years (*Texts 1–4*). . . . 65
- 4.3 The First Studies on the Parallel Postulate: 1677–1689 (*Texts 5–16*). . . . 70
- 4.4 Leibniz' Studies in the 1690s (*Texts 17–29* and *34*). . . . 80
- 4.5 Leibniz' Late Studies (*Texts 30–39*). . . . 91

5 **Reception and Legacy** . . . 103
- 5.1 Toward a Geometry of Space in the Eighteenth Century. . . . 103
- 5.2 Wolff, Kästner and the Early Reactions. . . . 105
- 5.3 The Concept of *Lage*, Direction, and the Actual Infinite. . . . 109
- 5.4 The Principle of Reason. . . . 115

6 **Leibniz' Texts on Parallel Lines**. . . . 125
- 6.1 From *Characteristica geometrica. Reading Notes to Fabri's "Synopsis geometrica"* (1673) . . . 125
- 6.2 From *De secandis parallelis* (1676). . . . 125
- 6.3 From *Linea infinita & extensio interminata* (January–April 1676) . . . 127
- 6.4 From *Dissertationis de arithmetica circuli quadratura propositiones septem and Quadraturae circuli arithmetica pars prima* (April–June 1676). . . . 129
- 6.5 From Leibniz' Marginal Notes to Barrow's Edition of Euclid (1676–1677). . . . 129

6.6 From *Characteristica geometrica* (January 1677) 129
6.7 From *Characteristica geometrica, Scheda 1* (1679) 129
6.8 From *Linea est via puncti ...* (1679) 131
6.9 *Axioma 13 Euclidis ...* (Around 1679) 131
6.10 From *De calculo algebraico et constructiones lineares optime conciliandis* (January 1680) 133
6.11 From *Elementa nova matheseos universalis* (Summer 1683) . . 135
6.12 From *Definitiones* (1685) 137
6.13 From *Si linea recta moveatur ...* (1685) 139
6.14 From *De curvis similibus et similiter positis et parallelis* (1685). 139
6.15 From *Logica de notionibus. Annotata circa schedas Jungianas* (1685) 141
6.16 From *Phoranomus* (1689). 143
6.17 From *Demonstrationes Euclideas* (Around 1690). 143
6.18 *Parallelae* (1690) 143
6.19 From *Generalia de natura linearum* (*Acta eruditorum*, September 1692). 145
6.20 *Rectae parallelae* (around 1692) 145
6.21 From *Lineam specie datam describere* (Around 1693) 147
6.22 From Leibniz' Notes to Arnauld's *Nouveaux elemens de geometrie* (Around 1693) 147
6.23 From *Elementa geometriae generalia* (Around 1694) 153
6.24 From *De novo usu centri gravitatis* (*Acta eruditorum*, November 1695). 153
6.25 From *Expediendi Laboris causa ...* (Around 1695) 153
6.26 From *Duae rectae parallelae ...* (1695). 155
6.27 Leibniz to Johann Bernoulli (January 29th/February 8th, 1697) . . 157
6.28 From *Specimen analyseos anagogicae* (1698) 159
6.29 From *Justification du Calcul des infinitesimales par celuy de l'Algebre ordinaire* (1702) 159
6.30 From *Attentius examinans ...* (Around 1702) 159
6.31 From *Calculum situs* (Around 1702) 161
6.32 From *Table de définitions* (1702–1704) 165
6.33 From *Nouveaux Essais sur l'entendement humain*, II, xxxi, § 3; and III, iii, § 18 (1703–1705). 165
6.34 From *De lineae super linea incessu* (*Acta eruditorum*, January 1706), and Letter to Johann Bernoulli (July 15th, 1706) . 167
6.35 From *Demonstratio omnimoda ...* (1712). 167
6.36 From *In Euclidis* πρῶτα (1712) 169
6.37 From *Spatium absolutum ...* (August–September 1714) 179
6.38 From *Rectam definio ...* (1715) 179
6.39 From *Calculi situs fundamenta ...* (1715). 179

Bibliography 181

Index of Names 191

I Leibniz on the Parallel Postulate and the Foundations of Geometry

1 Introduction

Among the many contributions of Gottfried Wilhelm Leibniz (1646–1716) to mathematics and philosophy, his work on the foundations of geometry is especially relevant. In Leibniz' times, the text of Euclid's *Elements* still represented the starting point for any advanced mathematical theory, including Leibniz' most celebrated discovery, the Calculus. The Greek treatise, on the other hand, was also the main model for deductive reasoning, and the touchstone of logical analysis and epistemology in general. In the seventeenth and eighteenth centuries, the debate on the *Elements* was extensive, and philosophers, philologists and mathematicians contributed, with dozens of emended and commented editions of the text, to a better understanding of Euclid's intentions and a deeper insight into the nature of geometry itself. Given Leibniz' great interest in logic, his involvement in foundational discussions about the new infinitesimal techniques, his wide erudition in the history of mathematics, and his didactical preoccupations with scientific education, it comes as no surprise that throughout his entire life he devoted a considerable part of his time to investigating the essence of geometrical reasoning or the system of principles needed to ground the whole of mathematics. Moreover, even though Leibniz often indulged in drawing *analogies* between his mathematical discoveries and his metaphysical views, it may be said that these two disciplines only *actually* merge together in his work on the foundations of geometry. In this field, Leibniz' most daring philosophical theses about the nature of space, his ambitious programs in logic and epistemology, and his mathematical expertise in dealing with Euclid's most celebrated theorems, simply combine into one science. He labeled his various researches on the topic under the general name of *analysis situs*.

Leibniz had several aims with this project, including the invention of a new formalism for geometry (the *characteristica geometrica*), which was intended to replace ordinary algebra in tackling geometrical problems and to constitute an important example of a more general *characteristica universalis*; the development of a geometry of *similarity* that could supersede or complement the common "geometry of magnitudes" with a non-metric theory leaning toward the geometry of perspective and projections; the foundation of his new infinitesimal Calculus on a sound geo-

V. De Risi, *Leibniz on the Parallel Postulate and the Foundations of Geometry*,
Science Networks. Historical Studies, DOI 10.1007/978-3-319-19863-7_1

metrical ground; the improvement of the science of mechanics with a new theory of space; and so forth.[1] One of the primary goals of Leibniz' geometrical studies, however, was to reshape, ground and bring rigor into the classical geometry of Euclid, by means of a thorough examination of the principles and proofs of the *Elements* in their ancient and modern editions.[2] The *analysis situs* is thus concerned with filling the gaps in the usual demonstrations of elementary geometry, that at the time were often relying on implicit assumptions or diagrammatic inferences, providing new and improved proofs of classical theorems that only employ explicit deductive steps and clearly showing the complete set of principles on which they depend. Moreover, these very principles (the axioms and postulates of the *Elements*, as well as a few other statements that were recognized as necessary to conclude a proof) were in turn discussed by Leibniz, who attempted to prove several of them and reduce the number of assumptions to a minimum. The ideal goal of *analysis situs*, in fact, was to prove *all* the axioms of geometry, and found it as a purely logical and unhypothetical science. Leaving aside, now, the epistemological grounds for such a grand and utopian aim, it is true that a large part of Leibniz' work on the foundations of geometry deals with the discussion of various Euclidean principles, their comparison, their mutual reduction, and finally (in some cases) their full demonstration.

Not surprisingly, several texts among Leibniz' essays on *analysis situs* deal with the famous Parallel Postulate. Since Antiquity, this principle was regarded as misplaced in a system of axioms (or postulates), and was almost universally considered as provable. Already at the time of Leibniz, several dozens of attempts had been made to prove it, which, however, resisted the efforts of the best mathematicians of the Early Modern Age untouched. The Postulate was seen as an ugly "blemish or mole on the beautiful body of geometry" (in the words of Henry Savile),[3] and its demonstration was included among the primary goals and concerns in the foundations of mathematics. Leibniz was perfectly aware of the regrettable situation of a simple geometrical statement which seemed impervious to proof, and attempted himself several demonstrations of the Parallel Postulate. It is quite clear that Leibniz was never satisfied with his proofs. He first tried to prove it, probably, in 1679; but in 1712, after more than thirty years of sporadic but untiring attempts, he had to admit to having attempted much, but proved nothing.[4]

---

[1] A comprehensive discussion of Leibniz' *analysis situs* is in De Risi 2007. Still useful, classical treatments of the subject are to be found in Couturat 1901 and Cassirer 1902. More recent papers are those by Freudenthal 1972, Münzenmayer 1979 and Wallwitz 1991. The edition of Leibniz' geometrical essays by Echeverría 1995 has a good introduction on the topic.

[2] On Leibniz and Euclid, see Knecht 1974, Giusti 1992, Echeverría 1992, Echeverría 1994, Bosinelli 1994. Besides his foundational studies, Leibniz sponsored a German edition of the *Elements*, which was published in 1696 by Meissner (see A iii, 7, nn. 59–60, pp. 225–30).

[3] This famous remark is taken from Savile's *Praelectiones tresdecim in principium elementorum Euclidis* (1621): "In pulcherrimo Geometriae corpore duo sunt naevi, duae labes, nec, quod sciam, plures, in quibus cluendis & emaculandis, cùm veterum tùm recentiorum, ut postea ostendam, vigilavit industria. Prior est hoc postulatum [the Parallel Postulate], posterior pertinet ad compositionem rationum" (lect. vii, p. 140).

[4] "Multa tentavi, et video ne hoc quidem facile demonstrari posse ...", *In Euclidis* πρῶτα, GM v, p. 202 (see below, *text 36*).

In any case, Leibniz never published anything on the topic, and his studies had no *direct* impact on the history of the attempts to prove the Parallel Postulate or the pre-history of non-Euclidean geometry. Some fragments of his work surfaced, if truth be told, in the nineteenth-century editions of Leibniz' writings; in particular, the all-important essay *In Euclidis* πρῶτα from 1712, which contains, among other things, a few demonstrations of the Parallel Postulate, was published by Gerhardt in 1858. In the second half of the nineteenth century (which marked the beginning of Leibniz' posthumous legacy as a source of inspiration for modern philosophy and science), Leibniz' theory of parallels was able to attract some quick reference in the foundational studies on geometry.[5] But these were soon forgotten, and today no history of mathematics (or non-Euclidean geometry) mentions Leibniz as the author of several proofs of the Parallel Postulate.

On the one hand, this is quite understandable. Since this part of Leibniz' work exerted no influence on the history of mathematics, and has no value in itself (since all his proofs are flawed, and their aim unattainable), it is reasonable to disregard it as a blind alley in the great man's manifold activities. Moreover, one has to admit that the degree of elaboration of Leibniz' fragments attempting to prove the Parallel Postulate often falls short of the best achievements of his contemporaries, and his attempts cannot be ranked among the *great* failures in the history of geometry.

On the other hand, these attempts are quite interesting in many respects, and surely deserve a better fate. They serve as an outstanding document of Leibniz' geometry, and shed much light on his project of an *analysis situs* in general. The plurality of techniques that Leibniz employs to prove the Parallel Postulate is impressive, ranging from purely symbolic reasoning to mechanical analogies, from mathematical projections to metaphysical reasoning, from differential geometry to a new theory of infinite wholes. These Leibnizian attempts largely surpass the imagination of his contemporaries, but also show some of the potentialities hidden in seventeenth-century mathematics. Since they are deeply entangled with Leibniz' epistemological and metaphysical views on the nature of geometry and space, they also constitute an important document for the understanding of Leibniz' philosophy.

I would say, however, that the relevance of these writings goes far beyond any particular interest for Leibniz' philosophy or the seventeenth-century history of

---

[5] The *In Euclidis* πρῶτα was published in the fifth volume of Gerhardt's edition of Leibniz' *Mathematische Schriften*. It was later cursorily quoted by Veronese 1891 (pp. 568–69), but only in relation to Proclus' proof of the Parallel Postulate and without taking into consideration Leibniz' own proofs. Leibniz' theory of parallels was also mentioned in Killing 1893 (vol. 1, p. 5), wrongly claiming that it was based on the concept of direction (Killing had probably in mind some nineteenth-century developments of Leibniz' *analysis situs*; see § 5.3 below). Finally, Hilbert, drawing on a German translation of Veronese, makes a reference to Leibniz and parallels in his lectures on the foundations of geometry from 1898–1899 (Hallett&Majer 2004, p. 263), although the quote didn't find its way to print in Hilbert 1899. Some years later, Frankland 1910 mentions Leibniz in relation to the Parallel Postulate (p. 19), but without any discussion of the matter. The only recent studies that deal with Leibniz' theory of parallels are the doctoral dissertation of Lauter 1953, the short article by Echeverría 1986 (which partially transcribes a Leibnizian paper on the topic, our *text 12*), and a forthcoming paper by Heis [a]. I thank Jeremy Heis for providing me with a draft of this article, as well as with another one on Kant's theory of parallels.

mathematics. The creation of non-Euclidean geometry in the nineteenth century, in fact, represents one of the greatest revolutions in the history of science, and a radical change in the significance and meaning of geometry itself. This mathematical revolution was prepared and made possible by a long transformation of the epistemological views about geometry, that took place in the previous century and in many respects stemmed from Leibniz' ideas. Leibniz' conception of space as a structure, and his new and all-important claim that geometry is the science of space (and thus a science of structures) provide the philosophical background needed to envisage the possibility of investigating different spatial structures and exploring alternative geometries. The enormous importance of the Parallel Postulate in this connection comes from the fact that it was considered the only principle of geometry still unproved and open to doubt, so that the only non-classical spaces considered at the beginning of the nineteenth century were described by a geometry in which *this* postulate is false. It is not surprising, then, that Leibniz' new conception of space as a mathematical structure comes especially to light in his papers on the Parallel Postulate. Yet, Leibniz (like all his contemporaries) was convinced of the truth and provability of the Postulate itself, and he envisaged a grand logicist project (much more radical than any other ever conceived by his contemporaries) devoted to putting Euclidean geometry on unshakable grounds. The tension between a new conception of space that would allow for the first time to sail in the unknown sea of modern geometries, and an epistemological program still anchored to Euclid's *Elements*, constitutes the major point of interest of Leibniz' papers on the theory of parallels. This tension forced Leibniz into the extreme attempt to reconcile geometry and epistemology bending them in unforeseen directions, mixing mathematics and philosophy, and eventually betraying the most basic assumptions of both disciplines. And even though Leibniz' fragments on the subject may lack the mathematical refinement and the richness in theorems of hyperbolic geometry that illuminate the later works by Saccheri and Lambert, his deeper and visionary insights on the nature of space powerfully contributed to the birth of modern geometry.

In the following pages, I first outline the historical background of Leibniz' studies on the Parallel Postulate (§ 2); then I try to sketch the most important features of Leibniz' epistemology of geometry and his program toward an *analysis situs* (§ 3). This is followed by an analytical presentation of the most relevant results from the collection of Leibniz' texts on the theory of parallels (§ 4). Finally, I turn to consider a few aspects of the poor reception of Leibniz' doctrines in the eighteenth century, since these (often misleading) interpretations were generally relevant for the history of non-Euclidean geometry (§ 5).

2 The Theory of Parallel Lines in the Age of Leibniz

2.1 From Antiquity to the Renaissance.

It may be useful to give a picture of the discussions on the Parallel Postulate in the age of Leibniz, as well as a list of his mathematical sources on the topic.[1]

The definition of parallel lines, in Euclid's wording, is that of "straight lines which, being in the same plane and being produced indefinitely in both directions, do not meet one another in either direction". The Parallel Postulate appears as Postulate Five in the First Book of the *Elements*, and states that "if a straight line falling on two straight lines make the interior angles on the same side less than two right angles, the two straight lines, if produced indefinitely, meet on the side on which are the angles less than the two right angles".[2] In Early Modern editions of Euclid, however, the Postulate was normally arranged among the *axioms*. Axioms were normally distinguished from postulates, meaning that the latter were principles employed in licensing geometrical constructions (e. g. Postulate One: "to draw a straight line from any point to any point") while the former expressed states of affairs (for instance: "two straight lines do not enclose a space").[3] In this systematization, the Parallel Postulate

[1] The classical reference for the history of non-Euclidean geometry is Bonola 1906, to which at least Pont 1986 should be added.

[2] Here and elsewhere, I take the English wording from Heath 1925.

[3] The classification of the Parallel Postulate among the axioms (or common notions) dates back at least to Geminus (first century BC), and was discussed by Proclus, *In Euclidis* 182–83. In modern times, Simon Grynaeus, the editor of the *editio princeps* of the *Elements* (1533), accepted Proclus' observation and classed (following a Greek manuscript he found) the Parallel Postulate as Axiom 11 (after the original Euclidean Common Notions and a few newly interpolated axioms). As most of the modern editions of Euclid were based on the Greek text established by Grynaeus, his classification of the principles was widely accepted, and the naming of the Parallel Postulate as Axiom 11 is very recurrent. Clavius adopts Grynaeus' numbering in the first edition of his commentary, but after having introduced one more axiom in his attempt to prove the Parallel Postulate (in the second edition of his commentary), he rearranges the order of the principles and the Parallel

V. De Risi, *Leibniz on the Parallel Postulate and the Foundations of Geometry*, Science Networks. Historical Studies, DOI 10.1007/978-3-319-19863-7_2

seems to fit better among the axioms, and Leibniz follows the Early Modern use in this respect. In any case, Leibniz' program to prove all the principles of geometry applies uniformly to both axioms and postulates.

Innumerable difficulties have been raised since Antiquity with regard to both the definition of parallels and the Parallel Postulate; with regard to the definition, because it refers to infinity, or at least to indefinite extension of straight lines; with regard to the postulate, as it seems too complex to be an authentic principle, and because (moreover) the inverse implication (*Elements* I, 28: "if a straight line falling on two straight lines make the interior angles on the same side equal to two right angles, the straight lines will be parallel to one another") is in fact provable without the need of any special principle.

In Late Antiquity, Proclus reported that several attempts had been made in the past to prove the Parallel Postulate, and he discussed some of them, eventually accepting one demonstration as sound. Roughly in the same era, the Aristotelian commentator Simplicius also transmitted a different proof of the same statement.[4] While Proclus' commentary was basically lost during the Middle Ages, Simplicius' work circulated in the Islamic world, engendering a long and beautiful tradition of attempts to prove the Parallel Postulate. The level of sophistication of these discussions was very high, and the works by Thābit ibn Qurra, Ibn al-Haytham, Omar Khayyām, and Nasīr ad-Dīn at-Tūsī on the theory of parallels introduced concepts and techniques which are among the most important developments in the history of the foundations of geometry.[5]

In the West, Simplicius was already known as early as the Middle Ages, and Proclus was rediscovered in the early sixteenth century; none of the more advanced Arabic sources, however, was available for a long time. The Parallel Postulate began to be widely discussed when the Jesuit mathematician Christoph Clavius (1538–1612) published his all-important edition of Euclid's *Elements* in 1574, including in it Proclus' proof of the Postulate and claiming it be a theorem.[6] Some years later, Clavius

---

Postulate becomes Axiom 13. In Leibniz' papers, it is sometimes called "Axiom 11", but more often "Axiom 13" as he was working mainly on Clavius' (later) edition. See De Risi [c].

[4] Proclus, *In Euclidis* 364–73. Simplicius *apud* an-Nayrizi, in Lo Bello 2003a, pp. 157–66 (original Arabic text in Besthorn&Heiberg 1932). On the latter text see Sabra 1969, who maintains that the proof appearing in an-Nayrizi is in fact by one Agapius ("Aganis" in the text; cf. Sabra 1968), while the original proof by Simplicius is to be found in another Arabic manuscript that Sabra edited (as well as in a couple of other Arabic texts).

[5] For a French translation of the most relevant Arabic texts, see Jaouiche 1986. Among the histories of non-Euclidean geometry, Rosenfeld 1988 is especially concerned with the developments in the Islamic World. As for the knowledge of medieval sources that Leibniz may have had, we are informed that several medieval Latin manuscripts were held in the library of Wolfenbüttel, that Leibniz directed for 25 years: see Folkerts 1981. A few of these Latin treatises were translated from Arabic, and were drawing on independent Arabic sources. These works, however, were limited to the most elementary developments of the rich season of Islamic medieval mathematics. Moreover, at the times of Leibniz all the information concerning these texts might be found in Renaissance editions of Euclid; we are thus not compelled to claim that Leibniz was actually reading Wolfenbüttel's manuscripts.

[6] Truth be told, the proof of Proclus was already published in the important Latin edition of the *Elements* by Federico Commandino, which appeared in 1572 (and in Italian translation in 1575).

came across an Arabic manuscript containing several versions of Nasīr ad-Dīn's commentary on the Parallel Postulate, and was able to extract some information from it. The text disclosed to him the rich and complex tradition of the geometrical researches of the Islamic world, and Clavius was forced to change his mind, disregard Proclus' demonstration of the Parallel Postulate as weak and flawed, and deeply revise Euclid's system of principles adding one more axiom and several demonstrations. He devised, then, two proofs of the Parallel Postulate, largely dependent on the Arabic sources, and published them in the second edition of his Euclidean commentary (1589).[7] Clavius' book was later re-issued several times and for two centuries it was considered the best and most complete edition of the *Elements* available to advanced scholarship, and the one text to be referred to on foundational issues in geometry. Thus, the entire discussion on the Parallel Postulate in the Early Modern Age started from, and continuously referred to, Clavius' work.

Clavius' first demonstration of the Postulate is especially relevant for us, as it relies on the concept of equidistance. Since Antiquity, in fact, various geometers had proposed to change Euclid's definition of parallels as non-incident straight lines into that of "equidistant straight lines". Starting with the latter definition, and implicitly assuming that the defined object (the equidistant straight lines) exists, it was possible to (correctly) prove the Parallel Postulate: and Proclus' demonstration was in fact relying on this definition and hidden existential assumption.[8] The mistake is

Clavius' 1574 edition is not very different on this issue, but it would improve after the second edition.

[7] In his commentary, Clavius opened the discussion of the Parallel Postulate acknowledging his Arabic sources: "Id quod in Euclide quodam Arabico factum etiam esse accepi, sed nunquam facta mihi est copia demonstrationem illam legendi, etsi obnixe illud iterum atque iterum ab eo, qui eum Euclidem Arabicum poßidet, flagitavi" (*Euclidis*, p. 50; in his copy of Clavius' book, Leibniz has underlined the passage). Nasīr ad-Dīn had written two different treatises on the Parallel Postulate (the second being a variation of the first), to which one should add a spurious essay, the so-called *Longer Version* edited by a pupil of his in 1298, a few years after Nasīr ad-Dīn's death. A copy of the three works was owned (along with many other Arabic manuscripts, such as that of Books 5–7 of Apollonius' *Conics*) by the Medici family; their librarian and orientalist Raimondi was appointed to publish most of these manuscripts, and to this end he founded and directed the Typographia Medicea in Rome. Raimondi published Nasīr ad-Dīn's *Longer Version* (in Arabic) in 1594. Nasīr ad-Dīn's works, however, had already been known in Rome for some years (the edition had started in 1588), and Clavius was therefore able to make some use of them in 1589, even though his relations with Raimondi were not excellent (see Baldini&Napolitani 1992). In fact, Clavius' proof in the *Euclidis* shows that he could also rely on the two original essays by Nasīr ad-Dīn, which are more complex and deeper than the published *Longer Version*. The three Arabic works are now published in French in the above-mentioned Jaouiche 1986. On the edition in Rome, see Knobloch 2002a, and (most of all) the volume edited by Cassinet 1986.

[8] This definition should date back to Posidonius (second century BC) and is reported by Heron, *Definitiones* 70, and Proclus, *In Euclidis* 176, who employed it in a proof of the Parallel Postulate in *In Euclidis* 371–73. It was discussed by Simplicius (or Agapius) and an-Nayrizi, too (see Tummers 1994, pp. 23–25, for the Latin text; cf. Lo Bello 2003b, pp. 39–41), who not only advanced a proof of the Parallel Postulate relying on equidistance, but also simply called "equidistant" the non-intersecting lines (as they assumed the truth of the Postulate). Since an-Nayrizi was well-known in the West from the Middle Ages, it happens that Renaissance and Early Modern editions of Euclid (like the one by Commandino) still called "equidistant" two straight lines that are defined as non-

that the definition assumes that the line equidistant to a straight line is itself straight. The assumption of the existence of equidistant *straight* lines is in fact equivalent to the Parallel Postulate, and in hyperbolic geometry (where the Postulate is false) the line equidistant to a straight line is not straight (it is a curve called *hypercycle*).[9] Clavius, who had accepted Proclus' proof in the first edition of his commentary, later realized that he had to prove that the equidistant line is straight, and tried to establish this result, first, by simply relying on a "good" definition of a straight line (which was itself one of the biggest foundational problems in the Early Modern Age, as Euclid's definition seems to be vague and obscure); and, second, with a kind of kinematic argument. This is as follows: let a straight segment make a right angle with a given straight line, and let it move along the line preserving the right angle (that is, it moves parallel to itself); then the flow of the other end of the segment describes a line which is equidistant to the given straight line. This equidistant line, moreover, has to be straight (this is the argument) as the motion is *uniform* and it always keeps the same direction; it seems impossible that such a flow could draw a curved line. This is false, of course, but the conclusion is almost irresistible for the imagination, and the mistake was hard to detect before a complete mathematical explanation of Gaussian curvature.[10] Clavius' argument from motion (which had been already employed by the Arabs, and by Thābit in particular) was repeated and modified innumerable times after him, and was still alluring geometers in the nineteenth century. We will see that some of the most important contributions of Leibniz to the foundations of geometry deal with the definition of a straight line, and with the exact, mathematical definition of uniformity, homogeneity, and other similar properties of space (and motion) that classical geometry had left vague. It will be clear, then, that any proof of the straightness of the equidistant line would be far from being elementary.

Clavius' second demonstration of the Parallel Postulate is less important for us (and for Leibniz). It is a simple variation on Nasīr ad-Dīn's original proof, and stumbles upon a wrong limiting process extending a local property to a global result.

intersecting. These editions then simply inadvertently skipped from the *name* of equidistance (that should characterize non-incident lines) to the *fact* of their actual equidistance, and thus were able to prove the Parallel Postulate through Proclus' demonstration (see COMMANDINO, *Euclidis*, p. 20). These texts contributed to blurring the fact that one has to prove the equidistance of non-intersecting straight lines.

[9] We may remark that the above-mentioned Heron (*Definitiones* 70) only defines equidistant *lines* (γραμμαί), and not equidistant *straight lines* (εὐθεῖαι); properly speaking, he is not making any mistake or surreptitious assumption. The *title* of his definition, however, explicitly mentions equidistant straight lines (but this title may be spurious) and his following definition of converging straight lines (*Definitiones* 71) also explicitly refers to εὐθεῖαι. We don't know whether Heron may have been aware of any foundational difficulty on this point. To my knowledge, the first mathematician to explicitly recognize the *petitio* is Alfonso de Valladolid (1270–1350), even though the repeated attempts of the Arab and Persian geometers to produce a proof of the Fifth Postulate not grounded on equidistance may reveal some awareness of the problem. It is clear, in any case, that the difficulty was not easily detected in the Renaissance and the Early Modern Age. On Alfonso's criticisms of the proof, see LÉVY 1992.

[10] For Clavius' proof, see the long *Scholium* before *Elements* I, 29, in his *Euclidis*, pp. 48–53.

Basically, it doesn't take into proper consideration the possibility of asymptotic lines, and thus fails to recognize the central issue in hyperbolic geometry. Men like Wallis and Leibniz, more used than Clavius (or Nasīr ad-Dīn) to dealing with the infinite in mathematics, would not fall into this trap, even though a complete and perspicuous explanation of Nasīr ad-Dīn's misstep would only be given by Saccheri.[11]

It is hard, in any case, to overemphasize the importance of Clavius' commentary for Leibniz. He studied Euclid's *Elements* from this work, and filled it with marginal notes. Almost all the references to Euclid in Leibniz' works are in fact references to Clavius, and it seems that he always read Euclid in Clavius' translation and accompanied by Clavius commentary. Leibniz' own foundational studies on elementary geometry often took the form of a literal commentary on Clavius' edition. Given the large time span in which these essays were written, we can imagine that Leibniz kept Clavius' book on his working desk for most of his life, and it surely represents the major source for his geometrical researches.[12]

After Clavius, the seventeenth-century discussion on the Parallel Postulate took several directions.

2.2 Jesuit and French Attempts in the Seventeenth Century.

The Jesuits went on to produce important commentaries on Euclid, but none of them reached the level of their common master. The greater part of these works were aimed at students of Jesuit colleges, and tried to simplify Clavius' presentation. It was common, for them, to present parallel lines as equidistant lines without any foundational concern; the Parallel Postulate was thus generally considered provable from this (flawed) definition. Among Jesuit textbooks, we have to mention the *Cursus mathematicus* (1661) by Kaspar Schott (1608–1666), which Leibniz read and annotated. Schott, a Jesuit disciple of Athanasius Kircher, defined parallel lines as non-incident straight lines (even though he immediately remarked that they are in fact equidistant) and thought to prove the Parallel Postulate just by noting that the inverse proposition (*Elements* I, 28) is provable – a line of reasoning that had been

[11] We know that Wallis was not satisfied with Nasīr ad-Dīn's proof, but he doesn't offer a refutation of it. In fact he could have been simply unsatisfied by the explicit assumptions made by the Persian mathematician, which may be not more evident than the Postulate itself. The spurious *Longer Version* of Nasīr ad-Dīn's commentary, as a matter of fact, assumed several unproved Lemmata, which Nasīr ad-Dīn had demonstrated in his original essays on the subject. It is true, however, that Wallis had access to the original essay as well (see below, *note* 26); in any case, he doesn't explicitly criticize any demonstrative step in Nasīr ad-Dīn's proof. See Wallis, *De postulato quinto*, p. 673. Saccheri's complete refutation of Nasīr ad-Dīn's proof is in the Third Scholium after Proposition 21 of *Euclides vindicatus* (pp. 38–41; see my notes *ad loc.* in De Risi 2014 [a]).

[12] Leibniz was using the 1607 edition of Clavius' commentary, which does not differ sensibly from the final 1612 edition (from which we quote). Leibniz' copy of Clavius' commentary is currently kept in the *Landesbibliothek* in Hannover, and has several handwritten marginal notes. Among them, we find a heavily annotated index, in which Leibniz spelled out the propositions needed to prove every theorem of Euclid, so to expose the deductive structure of the *Elements*.

followed, in the Renaissance, by Pierre de la Ramée, other Ramist logicians, and an otherwise fine geometer like Peletier.[13]

Among Jesuit texts that Leibniz knew, we should also include the important *Synopsis geometrica* (1669) by Honoré Fabri (1607–1688). This book, which raised a few interesting foundational issues (Fabri was a logician of some value), simply defined parallel lines as equidistant, and proved in passing the Parallel Postulate (as an uninteresting corollary). Similar stances were taken by André Tacquet (1612–1660) and Milliet Dechales (1621–1678) in their editions of the *Elements*.[14] The only exception to this trend in Jesuit schoolbooks is represented, of course, by Gerolamo Saccheri (1667–1733), who strongly criticized the many simplifications of his Jesuit fellows in the foundational domain, and aspiring to produce a thorough analysis of Euclid's axioms ended up with the first ever treatise on non-Euclidean geometry. Even though Leibniz knew about Saccheri and read (and criticized) his work on statics, however, the outstanding accomplishments on the theory of parallels of *Euclides vindicatus* (1733) arrived too late for him.[15]

---

[13] Peletier's edition of Euclid's *Elements* seems to be the source of Schott's "demonstration". As for Peletier, see his *Demonstrationum in Euclidis elementa*, p. 10. As for Schott, he defined parallel lines as non-incident and then as equidistant straights in these two statements: "34. Parallelae rectae lineae sunt, quae cum in eodem sint plano, et ex utraque parte in infinitum producantur, in neutram sibi mutuo incidunt. 35. Parallelogrammum est figura quadrilatera, cujus bina opposita latera sunt parallela, seu aequidistantia" (Schott, *Cursus*, Book 3, p. 64; cf. also Book 1, Chap. 3, art. 4, p. 6). He had already explained how to trace, in several ways, a parallel line to a given straight (*Cursus*, Book 1, Chap. 4, art. 2, pp. 9–10). And before proving *Elements* I, 29, he gave his proof of the Parallel Postulate based on the conversion of *Elements* I, 28. It goes as follows. *Elements* I, 28 has shown that if two straights are cut at right angles by a third line, they don't meet. It should follow, then, that if the sum of the angles formed by two straights with a transversal is less than π, then the two straight lines are "less divaricated" than earlier and thus they meet; should they not meet, in fact, they would be as divaricated as earlier, and thus would form two right angles with the transversal, which is absurd. Here the original of this very clumsy passage: "si enim duae rectae, AB, CD, tunc sunt parallelae, quando recta linea EF incidens facit angulos internos duobus rectis aequales, eo quod tunc a se invicem ita sunt divaricatae, ut concurrere non possint versus illas partes, ad quassunt recti anguli: ergo si facit eosdem duobus rectis minores, necessario concurrere debent ad partes, ubi minores sunt anguli, eo quod tunc minus a se invicem divaricentur quam antea: alioquin si aeque divaricarentur ut antea, recta linea EF incidens faceret duos angulos internos duobus rectis aequales ut antea. Adhiberi ergo in posterum potest Axioma illud 13. Euclidis, si non ut Axioma, saltem ut Theorema evidenter deductum ex demonstratione: tametsi id aliter et multis demonstrare conetur Proclus, Clavius, Tacquet, et alii" (*Cursus*, Book 3, p. 73). We may remark that the philosopher Salomon Maimon (1753–1800) was to advance a proof of the Parallel Postulate based on the conversion of *Elements* I, 28, as late as 1797; he grounded it on the Kantian claim that geometrical statements are synthetic *a priori* judgements, and on a further (non-Kantian) argument stating that in this class of propositions the subject and the predicate have to be coextesive: see the *Schlußanmerkung* to his *Kritische Untersuchungen über den menschlichen Geist* (pp. 361–70), as well as the commentary in Freudenthal 2006.

[14] Fabri's definition of parallels is in *Synopsis*, p. 30. His proof of the Parallel Postulate appears later as Corollary 3 to Proposition 1 (p. 83). Apart from reading Fabri's and Tacquet's works, Leibniz met Dechales in person when he was in Paris.

[15] Leibniz read Saccheri's *Neo-statica* (1708) and criticized it in a letter to Des Bosses dated August 23rd, 1713 (in GP ii, p. 482). During his lifetime, Saccheri enjoyed some celebrity in the Jesuit milieu, mostly because of his reputation as an extremely quick-minded thinker (he was a chess

In open rivalry with Jesuit textbooks, Antoine Arnauld (1612–1694) wrote a book on Euclidean geometry, the *Nouveaux elémens de géometrie* (first edition 1667; second revised edition 1683), which was meant to be the Jansenist treatise on mathematics, as well as (and more importantly) the Cartesian handbook on elementary geometry. Arnauld's attitude toward the Parallel Postulate is pretty equivocal, though. He defines parallel lines as non-incident lines, claiming that this is a "negative" definition, while the "positive" definition of parallelism is given through equidistance; the two definitions are then (rightly) regarded as equivalent by means of the Parallel Postulate.[16] As for proving the Postulate itself, Arnauld thinks that it is useless to try, since it is self-evident enough. In the end, however, he also gives Nasīr ad-Dīn's proof which he takes from Clavius (but without quoting him, as he was a Jesuit). Leibniz, who had an important correspondence with Arnauld on metaphysical topics, read and commented on the *Nouveaux elémens* in 1693.

Among the developments of the foundations of geometry in France, we should mention the geometrical work of Gilles Personne de Roberval (1602–1675), who already in 1642 was planning a new and improved version of the *Elements* (along the same lines as Arnauld). Roberval returned to the project in his old age, and was working on certain *Elemens de geometrie* in 1673–1675, the same years in which Leibniz joined the French community of mathematicians in Paris. On January 2nd, 1675, in a joint session of the *Académie des Sciences*, Leibniz presented to the scientific world his new calculating machine and Roberval his work on the foundations of geometry, which was intended to appear around Easter of the same year. In October, however, Roberval died without completing this book; and already in December Leibniz wrote to Oldenburg that he had been able to consult the manuscript. Roberval's *Elemens* were only published in the twentieth century, but Leibniz was greatly impressed by them and referred to Roberval's project throughout his life.[17] Roberval's aim was a complete reform of elementary geometry, as he was not satisfied by the carelessness of some Greek geometrical proofs, as well as by the system of principles governing Euclid's work. Roberval's theory of parallels did not offer any truly new solution to the problem, as he defined parallel lines as equidistant lines and then assumed an axiom stating that if two straight lines approach one another, they will continue to approach further on; a principle that had been already stated by Nasīr ad-Dīn and is

champion, for instance). Although he worked on the theory of parallels for more than thirty years, he didn't publish or anticipate anything on the subject and the *Euclides vindicatus* only went to print in the last weeks of his life.

[16] See ARNAULD, *Nouveaux elémens*, pp. 391–92 (ed. Descotes). The proof follows immediately thereafter.

[17] The edition of Roberval's *Elemens* is JULLIEN 1996, which also gives some details about the history of the manuscript. Leibniz' letter to Oldenburg dates from the 28th of December 1675 is in A III, 1, n. 70, p. 328, and is dismissive about the actual results of Roberval's work (he does not suggest the publication of the work). Either because Leibniz changed his opinion about Roberval's accomplishments, or because he nevertheless endorsed his epistemological program, in the following years he often mentioned Roberval's *Elemens* with great approval. For some quotes, see below *note* 7 in § 3.

in fact equivalent to the Parallel Postulate.[18] This notwithstanding, the *Elemens de geometrie* are of interest for our problem for several reasons. First of all, they expressed an epistemological and logical program for a reform of elementary geometry that was rarely stated with similar clarity before Leibniz. Second, their rigor in proving the results of the theory of parallels is truly remarkable, and many elementary results on quadrilaterals (proven without assuming the Parallel Postulate) are quite advanced and very relevant for the general theory (they would later be rediscovered by Saccheri). Third, Roberval had a fondness for purely logical reasoning that pushed him to prove simple results, such as the symmetry of the relation of parallelism, that was later to be shared by Leibniz as well.[19]

2.3 Italian and British Attempts in the Seventeenth Century.

Another line of development of the theory of parallels in the seventeenth century is to be found in the Italian school (consisting mostly in Galileo's legacy). Giovanni Alfonso Borelli (1608–1679) wrote an important *Euclides restitutus* (first edition 1658; third emended edition 1679), which is one of the most daring attempts in the foundations of geometry of the Early Modern Age. Borelli dealt mainly with the theory of proportions, but also with some minor issues in the theory of parallels. He believed that Euclid's definition of parallel lines is defective since it contains a reference to infinity, but recognized that the definition of parallels as equidistant

[18] Roberval's axiom, to be found in Jullien 1996, p. 201, rules out *ultraparallel* lines in hyperbolic geometry (straight lines that approach one another until they reach a minimal distance between them, and then diverge again to infinity) and is thus strong enough to force a Euclidean structure on the model. Roberval's general strategy in the Book IV of the *Elemens* is to prove a large number of results on quadrilaterals, assuming Euclidean hypotheses about them, and then prove through the above-mentioned axiom the truth of these hypotheses. The previous theorems, then, allow him to establish several Euclidean results (the first of which, in Proposition 19, is that two perpendiculars to the same straight line are equidistant – which is false in hyperbolic geometry) and finally prove the original Parallel Postulate in Proposition 27. (Note that Roberval proves *Elements* I, 29 as Proposition 17 without the need of his axiom, because he expresses the statement employing the concept of parallels as equidistant lines, and under this definition the proposition is in fact a theorem of absolute geometry).

[19] Roberval defines the distance between a point and a line through the length of the perpendicular from the point to the line. Then he says that two straight lines are *equidistant* if the distances from all the points of one of them to the other line are equal. Finally, he calls two lines *parallels* if one is equidistant from the other and the second is equidistant from the first. In Proposition 16, Roberval demonstrates the symmetry of the relation of equidistance, thus proving (in his terminology) that equidistant lines are also necessarily parallels. The proof, which does not require further principles (nor the Parallel Postulate), is relevant because it opens the way to the consideration of parallelism as an equivalence relation, that was to interest Leibniz in his formal essays on *characteristica geometrica* and was to represent, later on, an important development in the attempts to prove the Parallel Postulate. Kant himself, in his attempt to prove the Postulate, was to insist much on the symmetry of the relation of equidistance (see his *Reflexionen* 8–10, in KgS xiv, pp. 33–51). The antecedent of Roberval's proof may be found in Thābit ibn Qurra, even though the two treatises on parallel lines by the great Arab mathematician were probably unknown to both Roberval and Leibniz (see Rashed&Houzel 2005).

lines is faulty. Thus, he defined parallel lines as straight lines which have a common perpendicular, and then stated the Parallel Postulate (which he did not try to prove) saying that parallel lines (so defined) are equidistant.[20] Borelli's fine argumentative turn, which dispensed with the attempts to prove the Postulate, was followed by Angelo Marchetti (1674–1753), whose *Euclides reformatus* (1709) accepted both Borelli's definition of parallel lines and the new formulation of the Postulate.[21] It was not followed by others, though. Vitale Giordano da Bitonto (1633–1711) wrote in 1680 (second edition 1686) an *Euclide restituto*, in Italian, which was intended as a posthumous homage to Borelli. Giordano saw with great clarity all the difficulties involved in proving the Parallel Postulate (the paralogisms deriving from the notion of equidistance, or the uniform motion, etc.) and accepted a definition of parallels as equidistant lines only to say that their actual existence should be proved through the Euclidean Postulate, and thus attempted a new and original demonstration of it (which takes up many pages of his volume). In the end, he also stumbled upon a limiting process, and claimed to have proven the famous statement; however, his attempt was complex and elegant, and is considered one of the best constructions before Saccheri.[22] Leibniz was well acquainted with these developments. He had an interesting correspondence with Giordano on the foundations of geometry (even though not on the theory of parallels properly speaking) in 1689, and was able to

[20] Borelli introduced his new definition of parallel lines because it refers to a most clear and evident property ("cum exponatur per passionem possibilem & evidentissimam", as we read in the *Preface* of his *Euclides*). Moreover, since Euclid is able to construct the perpendicular to a given straight line without resorting to the Parallel Postulate (by *Elements* I, 10 and 11), one is able to show that Borelli's parallels exist without appealing to the Postulate itself (as Euclid does, apparently, in his *Elements* I, 31, at least to guarantee the uniqueness of the construction), or to another axiom about equidistant lines. Borelli's epistemological concerns about definitions, thus, are very similar to those by Leibniz about the "real definition" of a geometrical object (see below § 3). Borelli's formulation of the Parallel Postulate, in fact, directly assumes what Clavius had claimed to prove with his clumsy reference to the uniformity of motion, i. e. that the movement of a segment perpendicular to a straight line traces another straight line: "Axioma XIV. Si recta linea, in suo extremo semper perpendiculariter constituta super aliam rectam lineam, moveatur in transversum in eodem plano: alterum punctum extremum translatae rectae lineae in eius fluxu rectam lineam describet" (*Euclides restitutus*, 1658, p. 32; as Axiom 13 in the 1679 edition, p. 23). To this axiom, Leibniz remarked (in his copy of Borelli's book) that there is no necessity to assume that the flowing line is perpendicular to the given straight.

[21] See Marchetti, *Euclides reformatus*, pp. 10–11.

[22] Giordano criticisms of the Euclidean definition of parallel lines follow Borelli's own qualms about the use of infinity in a definition (see *Euclide restituto*, pp. 12–13). His demonstration of the Parallel Postulate occurs just before *Elements* I, 29, and can be read in *Euclide restituto*, pp. 60–66. Giordano's attempt in this direction is especially relevant because he discusses at some length the properties of the so-called "Saccheri quadrilaterals". Even though these figures (quadrilaterals with two equal sides at right angles with the base) were already treated in Medieval proofs (especially in Khayyām) and in Clavius, no one before Saccheri saw their foundational importance as Giordano did. On this point see Bonola 1905. It may be remarked that the last demonstration of the Parallel Postulate attempted by Leibniz (in the *In Euclidis* πρῶτα) also discusses these quadrilaterals (the Parallel Postulate being proved if one can show that the opposite angles in a Saccheri quadrilateral are also right angles; see *text 36* below).

read and annotate Giordano's *Euclide*.[23] By 1678, moreover, Leibniz had acquired Borelli's book (from the library of Martin Fogel), and Giordano recommended that he read it attentively; we have, in fact, Leibniz' marginalia to Borelli's *Euclides* as well.[24] Later in his life, Leibniz also had an exchange with Marchetti, but apparently they did not correspond on the theory of parallels.[25]

A further important line of development of the theory of parallels in the seventeenth century was engendered by the above-mentioned opinion of Henry Savile (1549–1622) that the Parallel Postulate is an ugly *naevus* in Euclid. Decades later, John Wallis (1613–1703), who was appointed Savillian professor of geometry in Oxford, decided to honor the will of his benefactor and directed his genius to the Euclidean blemish. In 1651 Wallis had Nasīr ad-Dīn's book translated from Arabic into Latin by his colleague Edward Pocock (thus reading for the first time the original Medieval treatise instead of Clavius' interpretation of it), and in 1663 he himself

[23] Leibniz read Giordano's *Euclide* (in the 1686 edition) while he was traveling in Rome, and wrote to him shortly thereafter. The whole correspondence amounts to only three letters (two by Leibniz and one by Giordano), which are however very rich in epistemological and geometrical remarks. They were first published in GM I, pp. 195–200, but the critical edition in A III, 4, nn. 216–18, pp. 420–28, is also useful as Leibniz' draft of the second letter to Giordano is much clearer than the letter he actually sent. The correspondence is mostly concerned with providing a good definition of a straight line, and touches in passing the definition of parallels. As for Leibniz' marginalia to Giordano's book, they only concern Proclus' proof (reported by Giordano on p. 22 of his *Euclide*) that two straight lines cannot have any common segment. To this proof, Leibniz objects that it should be extended taking into consideration that two straight lines might have even more than one segment in common (while Proclus' proof is based on the fact that, *per absurdum*, they might have one; and it only works in this restricted hypothesis). Leibniz expressed his dissatisfaction about this proof in his first letter to Giordano. As Leibniz' marginal note in Giordano's *Euclide* is still unpublished, I transcribe it here (it is in Italian, like Giordano's book): "Si suppone che G et F sono punti differenti, il che non è necessario, perché non è dimostrato ancora che le due rette AD, CD, non possono concorrere nel punto G overo F dove il circolo le sega. Questo difetto è perdonabile, perché al meno non tutti [i] puncti possono essere li mede[si]mi, dove un altro circolo descritto dal centro B sega queste rette altrimenti non si darebbero …". Leibniz adds at the beginning of p. 24 (Proclus' proof that two straight lines do not enclose a space) a similar remark: "Si suppone ancora che siano diversi". There is nothing about Giordano's proof of the Parallel Postulate.

[24] We may also mention that Michelangelo Fardella had been a pupil of Borelli, and Leibniz met him during his trip to Italy in 1690 (at the same time in which he met Giordano), and later had an important correspondence with him (see GARBER 2004). Leibniz' reading of Borelli's *Euclides* (or better a re-reading of it) may have been very late (around 1712): see DE RISI 2007, pp. 119–20. Among the many marginalia in this book, we may mention a few about the concept of dimension (DE RISI 2007, pp. 207–208), a couple concerning the definition of a straight line, and two complaining that Borelli (correctly, we would say) assumed so many unproved axioms in his book ("hoc demonstrandum"!). Leibniz' final judgement on Borelli was however quite dismissive: "In Borello multum diligentiae agnosco, ingenio mediocri fuisse videtur" (Leibniz to Johann Bernoulli, October 23rd, 1716; in GM III, p. 971).

[25] Leibniz had a brief exchange with Marchetti in the 1690s (A III, 6, n. 244, pp. 806–807; A III, 7, n. 196, pp. 779–81) about the latter's new theory of proportion published in 1695. Here, Leibniz approves the foundational stance of Marchetti and Borelli before him, only regretting that the analysis of the principles was not complete (A III, 7, p. 780). Marchetti's book on Euclid (dealing with the theory of parallels) was only published some fifteen years later, when apparently Leibniz was no longer in contact with him (cf. BODEMANN 1889, p. 167).

attempted a demonstration of the Postulate. Both essays (Wallis' and Nasīr ad-Dīn's) were later (1693) published in an appendix to Wallis' treatise on algebra.[26] Wallis' proof is of the utmost importance, as it shows that the Parallel Postulate is in fact equivalent to the possibility of transformations through similarity. This means that in non-Euclidean spaces there are no (non-trivial) similar figures; for instance, any two triangles with the same angles are in fact isometric (congruent) with one another. Wallis considered this result to be not so much a characterization of the Parallel Postulate (an alternative formulation of it), as an actual proof, as he could not conceive that the possibility of similarities could be denied. The reasoning behind this last claim seems to be (besides an appeal to the principle of continuity) that similarities preserve the quality (i.e. the shape) of the figures, while they do not preserve their quantity (i.e. length, area or volume); but since quality and quantity are (in Aristotle) different categories of being, it must be possible to change a feature in a figure pertaining to one of them while leaving the other untouched (i.e. change the quantity while preserving the quality).[27] Wallis' proof convinced few mathematicians, and already Saccheri complained that his demonstration of the Parallel Postulate was more metaphysical than mathematical, but it was widely discussed by philosophers up to the nineteenth century.[28] In 1696, Leibniz wrote a review of Wallis' *Algebra*, mentioning the appendix, but he didn't enter into the philosophical discussion about similarity.[29] In his writings on *analysis situs*, however, Leibniz developed a vast theory of similarity which he considered to be the cornerstone of his new geometry; a

[26] Wallis' treatise of algebra appeared in English in 1685, and in Latin (with some appendices) in 1693. In the latter form, it was printed as the second volume of Wallis' *Opera mathematica*, and included the short essay from 1663 entitled *De Postulato Quinto et Definitione Quinta Libri 6 Euclidis Disceptatio geometrica* at pp. 665–78 (pp. 669–73 being Pocock's Latin translation of Nasīr ad-Dīn). We know that Pocock translated into Latin both the original work of Nasīr ad-Dīn and the spurious essay that was published in Rome. Wallis' *Algebra* only contains the latter, but Wallis himself had access to Pocock's translation of the former, which he added in manuscript to his own copy of the *Opera mathematica*. Pocock's unpublished translation is still available in manuscript at the Bodleian Library, and was reprinted in Cassinet 1986.

[27] *De postulato quinto*, Lemma 8, p. 676. But cf. also Wallis' *Institutio logicae*, in *Opera*, vol. 3 ii, p. 105 (on the concept of figure as quality).

[28] We know Saccheri's opinion on Wallis' proof from a letter sent on the 12th of July 1713 by Tommaso Ceva to Guido Grandi, in which Ceva reports Saccheri saying: "… quantunque il suo lemma ottavo resti da lui provato più tosto con la metafisica che con rigore geometrico, onde pare che fosse necessaria la costruzione problematica per torre ogni sospetto di petizione di principio" (in Tenca 1952, p. 35). The opinion was probably shared by many other mathematicians, as no one was prevented by Wallis' alleged proof from attempting to demonstrate the Parallel Postulate. A similar position was later held by Lambert, *Theorie der Parallellinien*, §§ 79–81 (pp. 350–2), who however was willing to accept Wallis' principle on similarity as an unprovable axiom.

[29] Leibniz' review of Wallis appeared in the 1696 issue of the *Acta Eruditorum* (pp. 249–59). We have also several unpublished notes on Wallis' volume and drafts of the review: see LH XXXV, vii, 22; and cf. LH XXXV, xiv, 2, Bl. 66–75 and 90. Among these papers, however, there is not much about Wallis' proof of the Postulate, as Leibniz preferred to concentrate on the algebraic results proper. In the published review, the whole appendix on the Parallel Postulate is disposed of with a dismissive note: "Peculiari etiam dissertatione defenduntur definitio quinta & postulatum quintum libri sexti Euclidis, quae Savilius ipse inter propositiones demonstrabiles rectius referri in suis Lectionibus judicaverat" (p. 259). As even the quotation is wrong (the Parallel Postulate

theory, moreover, that had important consequences in epistemology and metaphysics. Thus, Wallis may have at least convinced him that a proof of the Parallel Postulate was needed not only to fill a gap in the system of principles of elementary geometry, but also to give a foundation to his own philosophical projects.[30]

Lastly, we should mention a clumsy move in the development of the theory of parallels, by Thomas Hobbes (1588–1679). In his influential work on natural philosophy from 1655, the treatise *De corpore*, Hobbes advanced several mathematical theses that were in turn attacked by Wallis himself; Hobbes answered with *Six lessons to the Savillian professors of mathematics* that appeared as an appendix to the 1656 English edition of the same treatise. In these essays, which predated Wallis' attempts to prove the Postulate and in fact even Borelli's criticisms of the *Elements*, Hobbes claimed that the Euclidean definition of parallel lines is obscure and that a sound mathematical definition should show the way to generate its object (which the Euclidean definition does not). Later on, however, he gave a non-constructive definition of parallels, saying that they are equidistant lines, that is, lines "between which every line drawn, in any angle, is equal to any other line drawn in the same angle". After claiming that this definition is quite general and good for both straight and curved lines (and surfaces), Hobbes simply assumed that one can find straight lines behaving in this way, and thus (making the old mistake of Proclus) proved the Parallel Postulate from a definition that would need itself need a proof. Hobbes concluded that he had "done that business for which Dr. Wallis receives the wages"

being the Fifth Postulate of Book One, not Book Six), one suspects that Leibniz didn't grasp the full importance of Wallis' proof at the time.

[30] Among Borelli's reflections on the role of principles in geometry (in the first edition of his *Euclides restitutus*, pp. 14–17), we find a discussion on the necessity of proving the existence of a defined mathematical object, as for instance the parallel lines in his own definition (see above at *note* 20 his reference to *Elements* I, 10 and 11 to this effect). It is very remarkable that Borelli's examples of definitions, whose objects are doubtful before an existential or constructive proof, are: parallel lines (in Euclid's definition), squares, similar figures. There is no further textual hint that he connected these various examples under the same label. Since however he must have tried hard to prove the Parallel Postulate and surely read the extant literature on the subject, he could not have failed to notice that the theory of parallels was implied in the possibility of constructing a square (as it is clear in Clavius and in the Arabs through the use of Saccheri quadrilaterals). It is not unlikely, then, that Borelli also realized that the Parallel Postulate is needed to ground the possibility of similar figures. If this is true, he may have conceived Wallis' "proof" of the Postulate and disregarded it as too philosophical (one should also bear in mind that a sketch of Wallis' proof may be recognized in Clavius: cf. Maierù 1978; and possibly even in Euclid, *Data* 34–38). Be that as it may, Leibniz (who seemed unreceptive to Wallis' own demonstration) carefully underlined Borelli's sentence on his copy of *Euclides restitutus*. Here is the original quote: "Deinde si constructio aut passio nominata, sit quidem possibilis, & vera, sed nobis ignota, aut dubia; tunc bona definitio non erit: Nam conclusiones ab ignoto, & dubio principio ortae, incertae quoque, & dubiae erunt; & ideo suspicionem, aut opinionem, non autem scientiam certam afferent. Ut cùm dicitur in vulgata definitione parallelarum: Duae rectae lineae in eodem plano ex utraque parte non concurrentes, vocetur parallelae. Ignoratur an dari possint duae rectae lineae, habentes hanc conditionem. Similiter ignoratur, an reperiri possit in natura figura quadrilatera, in qua omnes anguli sint recti, & quatuor latera sint inter se aequalia, quae vocatur quadratum. Pari modo incertum est an reperiri possint figurae planae aequiangulae, habentes circa angulos aequales latera proportionalia, quae similes vocantur" (*Euclides restitutus*, p. 17).

as the Savillian Professor of mathematics.[31] Even though Hobbes' proof is so badly conceived as not to have any place in the history of the attempts to prove the Postulate, Leibniz knew Hobbes' mathematical books very well, and was influenced by Hobbes' attempt to give a general definition of parallel lines that could be applied to both straight and curved lines. Most of all, Hobbes' theory of constructive definitions (forsaken by Hobbes himself) was quite important in Leibniz' epistemology of mathematics, and this, in turn, heavily affected his geometrical studies.

[31] On the need for constructive definitions, see for instance the *Six lessons*: "And where there is place for demonstration, if the first principles, that is to say, the definitions contain not the generation of the subject, there can be nothing demonstrated as it ought to be" (*The works*, vol. 7, p. 184). A similar passage is to be found in *De corpore* I, VI, § 13, but the reason Hobbes provides for the need for causal definitions is here to be found in the Aristotelian idea that scientific knowledge is causal knowledge (the διότι), rather than in the ontological doubt about the real possibility of an object. (see below § 3.2). Hobbes' original definition of parallels, in Latin, is in *De corpore*, II, XIV, § 12, and with it Hobbes proves *Elements* I, 29 (equivalent to the Parallel Postulate) as a Corollary, as well as the theorem that the interior angle sum of a triangle is π (*Opera*, vol. 1, pp. 163–65). The definition and the proofs were correctly criticized by Wallis in his *Elenchus geometriae hobbianae* (pp. 29–32), explaining several difficulties but not pointing out the main flaw (that Hobbes had to prove that a parallel to a straight line is straight as well), and Hobbes could re-state his original definition and demonstration (with only a small and insignificant correction) in the English edition of the *De corpore* (*The works*, vol. 1, pp. 189–91; where he also mentions similar criticisms raised by Mylon) and the *Six lessons*, in which he also better explains the alleged shortcomings of the Euclidean definition (*The works*, vol. 7, pp. 205–207, from which my quote is taken; and again pp. 254–55). Hobbes' claim to have settled Savile's problem better than Wallis is again in the *Six lessons* (*The works*, vol. 7, p. 185). On Hobbes' long and savage dispute with Wallis, see JESSEPH 1999 and BEELEY&PROBST 2005.

Leibniz' Epistemology of Geometry and the Parallel Postulate

3

3.1 Proving Axioms.

The seventeenth-century mathematicians whom we have mentioned so far did not share a common epistemology, and each of them had his own opinions about the nature, object and aims of geometry, as well as on the meaning of geometrical principles or the standards of rigor needed in a proof. This notwithstanding, there was some kind of consensus at least on this point: that the axioms of geometry were provable. Contrary to our modern understanding of the role of an axiom as an indemonstrable statement, they claimed that geometrical principles were self-evident truths that *required* no proof to be accepted, but nevertheless were not *incapable* of proof. The dispute was more about the *opportunity* to prove them than the possibility of doing this.[1]

[1] I don't claim that this general consensus was truly universal among mathematicians and philosophers. But even though some authors may have maintained that mathematical axioms are indemonstrable in the strictest sense (appealing to a possible reading, for instance, of Aristotle's theory of principles in the *Posterior Analytics*), I would say that the largest part of the mathematicians and epistemologists known to Leibniz claimed the contrary. One should also consider that some geometers said an axiom is "indemonstrable" if it follows so immediately from the definitions, that there isn't logical space enough, so to speak, to articulate a proof of it; this notwithstanding, the axiom is clearly derivative from the definitions and not an assumption in itself. See for instance: "Sunt axiomata, ut dictum est, propositiones speculativae, seu theoremata, quae ob sui maximam evidentiam non indigent demonstratione; sed statim, cognitis terminis, neceße est, ut intellectus illis aßentiatur; & propterea aßumuntur inter principia indemonstrabilia" (Borelli, *Euclides restitutus*, p. 14). The same is stated by Hobbes: "As for those *notiones communes*, called *axioms*, they are from the definitions of their terms demonstrable, though they be so evident as they need not demonstration" (*Six lessons*, in *The works*, vol. 7, p. 210; cf. also pp. 199–200, and *De corpore*, II, VI, § 13, in *Opera*, vol. 1, p. 71–73). See also Schott about the second and third kind of principles (the first one being the definitions themselves): "In secundo genere sunt Petitiones quaedam, seu Postulata; quae quidem adeo clara sunt, ut tametsi demonstrari possint, nullâ tamen indigeant confirmatione, sed auditoris tantum assensum deposcant, ne ulla sit in demonstrando haesitatio, aut difficultas. In tertio denique genere sunt Axiomata, seu communes animi notiones

V. De Risi, *Leibniz on the Parallel Postulate and the Foundations of Geometry*,
Science Networks. Historical Studies, DOI 10.1007/978-3-319-19863-7_3

Clavius, on the one side of the epistemological spectrum, added a demonstration to every axiom in the *Elements*. These demonstrations of the axioms depend on the definitions of the mathematical terms, and the underlying epistemology seems to be that a statement whose proof follows *immediately* (whatever this means) from definitions alone is an *axiom*, whereas a statement that requires further means in order to be proved (other axioms, constructions, theorems) is a *theorem* in the proper sense. For instance, that only one straight line passes between two points is an axiom because it follows (or should follow) from the definition of a straight line; whereas Pythagoras' Theorem is proven through ruler and compass constructions and the help of several other axioms and theorems, and thus counts as a theorem. The same happens with the Parallel Postulate itself, that Clavius proves after *Elements* I, 28. From this point of view, the self-evidence that characterizes an axiom is not a reason not to demonstrate it, but a *consequence* of the immediateness of its proof.[2]

On the other side of the epistemological divide, Arnauld and the Cartesian school claimed that the self-evidence of the axioms makes the proof of them useless. To prove them, they said, one has to have recourse to principles and statements that are less evident then the axioms themselves, thus defeating the purpose of the proof,

(Effata, seu Pronuntiata alij appellant) quae non solum in Mathematica scientia, sed etiam in omnibus alijs adeo manifesta sunt, ut ab illis dissentire nequeat qui vocabulorum significata percipit" (*Cursus*, Book 3, p. 62). Among Leibniz' correspondents, Hermann Conring was in fact surprised by an epistemological program purporting to prove the axioms of geometry, and he claimed that no one had attempted this before, nor had conceived axioms as provable: "Demonstrationem *nihil aliud esse quam catenam definitionum*, nemo quod sciam alius asseruit. Nec forte etiam Tu serio hoc assertum velis. ... Quod adfers pro analysi Mathematicorum, reapse novum est commentum tuum. Mihi res fuit cum Mathematicis priscis, qui tuum illud non habent. Omnia axiomata esse demonstrabilia, novum plane est pronunciatum. Mea assertio stat firmis subnixa fundamentis et ab omni aevo hactenus fuit probata. Tuam ipse destruis dum ais: *Omnes propositiones certae demonstrari possunt praeter identicas et empiricas*" (Conring to Leibniz, February 28, 1678; in A II, 1, n. 167, pp. 595–596). Leibniz answered that, on the contrary, this was the common opinion of the Schoolmen: "Idque Scholastici omnes confirmant, qui axiomatum veritatem patere ajunt intellectis terminis; id est posse ipsa facili negotio nec longa definitionum serie demonstrari, sive ad identica reduci; sive quod idem est semper ostendi posse, quod contrarium implicet contradictionem" (Leibniz to Conring, March 19th, 1678; in A II, 1, n. 168, p. 599; see below, *note* 19). Similar references to an established tradition appear in this connection also elsewhere; see *Nouveaux Essais*, IV, VII, § 1 (A VI, 6, n. 2, p. 407, quoted later in *note* 7). Johann Bernoulli was perplexed over the issue, too, and he stated that without assuming some axioms the demonstration would be a *regressus in infinitum* (letter to Leibniz from August 25th, 1696; in A III, 7, n. 27, p. 102). Leibniz' answer, very confident in restating his epistemological opinions, is quoted below in *note* 7.

[2] It may be hard to exactly spell out the conditions that make a proof *immediate* in Clavius' sense. To be sure, he tried to keep the distinction between axioms and theorems intact through this vague notion of immediateness, and claimed that all the Euclidean axioms are properly axioms even though they are provable, with the only exception being the Parallel Postulate, whose proof is so complex that it deserves the name of "theorem" (*Euclidis*, p. 53). Other geometers may have devised different and more complex epistemological stances; for instance, Saccheri would like to claim that the (proven) Parallel Postulate is in fact an axiom and is not transformed into a theorem through his own demonstration, the reason being that he used a special logical rule (the *consequentia mirabilis*) to carry out the proof (see my *Introduction* in De Risi 2014 [A]). Leibniz simply held that all the usual Euclidean axioms are in fact theorems: "Itaque axiomata illa vix aliud erunt quam theoremata valde clara" (A II, 1, n. 218, p. 771).

which is to provide (psychological) grounds for accepting the statement. Thus, for instance, Arnauld thinks that it is not wise to "lose time and rack one's brain" to prove the Parallel Postulate, which has "clarity enough" to be accepted without demonstration.[3] And Malebranche mocks the pedantic Savile for having wasted his time with such futilities.[4] In this connection, the deduction of an axiom from the definitions should not be considered as a proof of it, since this kind of logical derivation does not clarify the axiom for us.

The point was whether a "demonstration" was to be intended as a logical deduction or a rhetorical tool; but both Cartesians and Schoolmen agreed that axioms are provable in the former sense of the word, as they are logically derivable from definitions. We have seen, in fact, that in his geometrical work Arnauld is willing enough "*de se rompre la tête*" to prove (i. e. logically derive) the Parallel Postulate from the definition of parallel lines using Nasīr ad-Dīn's proof, even though he cannot admit that this would amount to a true and fully fledged (rhetorical) demonstration.[5]

---

[3] See the general statement in the *Logique de Port-Royal*: "II Defaut. Prouver des choses qui n'ont pas besoin de preuves. … C'est ce qui a porté Euclide à prouver que les deux côtés d'un triangle pris ensemble sont plus grands qu'un seul, quoique cela soit évident par la seule notion de la ligne droite, qui est la plus courte longueur qui puisse se donner entre deux points, & la mesure naturelle de la distance d'un point à un point, ce qu'elle ne seroit pas, si elle n'étoit aussi la plus courte de toutes les lignes qui puissent être tirées d'un point à un point" (ARNAULD & NICOLE, *La logique ou l'art de penser*, IV, 9, pp. 326–27). On the Parallel Postulate properly: "Sixième Axiome. Deux lignes droites qui étant prolongées vers un même côté s'approchent peu à peu, se couperont à la fin. *Euclide prend cette proposition pour un principe et avec raison: car elle a assez de clarté pour s'en contenter, et ce serait perdre le temps inutilement que de se rompre la tête pour le prouver par un long circuit*" (ARNAULD, *Nouveaux elémens*, p. 361; ed. Descotes).

[4] See MALEBRANCHE, *Recherche de la vérité*, II, II, 6 (in *Oeuvres*, vol. 1, pp. 297–301).

[5] Among other Cartesians adopting the same point of view, see at least the grand *Cours entier de philosophie* (1691) by Régis, who gives the (psychological) distinction between demonstrating a theorem and explaining an axiom in his *Logique*, II, 3 (*Cours*, vol. 1, pp. 37–41), and then again treating the *lumen naturale* in his *Metaphysique* II, I, 19 (pp. 382–89). At the beginning of this tradition of thought, we may also include Pascal, who in *De l'esprit géometrique* clearly states that geometrical definitions are the very foundations of this science, and nevertheless one has to accept unproved axioms: "Aussy, en poussant les recherches de plus en plus, on arrive necessairement à des mots primitifs qu'on ne peut plus definer, et à des principes si clairs qu'on n'en trouve plus qui le soient davantage pour servir à leur prevue. D'où il paroist que les hommes sont dans une impuissance naturelle et immuable de traiter quelque science que ce soit dans un ordre absolument accomply. … Car, quand elle [*viz.* geometry] est arrivée aux premières veritez connues, elle s'arreste là et demande qu'on les accorde, n'ayant rien de plus clair pour les prouver: de sorte que tout ce que la geometrie propose est parfaitement demonstré, ou par la lumière naturelle, ou par les preuves" (PASCAL, *Oeuvres*, vol. 9, pp. 246 and 254). A few fragments of this text were already known at the time of Leibniz, since they had been published in Port Royal *Logique* (cf. here the first *Discours*, p. 21); but Leibniz had been able to look at Pascal's manuscript in its entirety when he was in Paris (see Leibniz' *Projet et essais*, in A VI, 4A, n. 205, p. 970). Leibniz praised Pascal's aim to prove any truth that was not completely clear in geometry, deliberately ignoring the fact that Pascal also claimed that axioms were in fact unprovable: see Leibniz' *Meditationes de cognitione, veritate et ideis* (in A VI, 4A, n. 141, p. 591). Elsewhere, Leibniz recognized that Pascal had denied the possibility of proving the principles, and tried to show why he was wrong (see below the quote from the *Paraenesis*, in *note* 18).

It is important to understand this general trend in the seventeenth-century epistemology of mathematics, namely, one toward assuming that all the axioms are provable (logically deducible) and that geometry thus ultimately rests on definitions only, which are the true principles of demonstration (in the formal sense). The status of definitions themselves was often unclear or debated. In practice, they acted as "axioms in disguise" in most mathematical proofs. This was the case, for instance, of the above-mentioned definition of parallels as equidistant straight lines (a definition equivalent to the Parallel Postulate, at least if one assumes the existence of such lines); but other definitions (such as that of a straight line) also had such a hidden axiomatic character, and it was only thanks to them that geometers and philosophers were able to dream of geometry as an unhypothetical and axiomless science.

Leibniz' epistemology of mathematics and his "logicist" program in geometry are thus to be framed against this common background. He was convinced that the axioms can be proved starting from the definitions, and between Clavius and Arnauld he decisively sided with the former. Leibniz thought that the notion of a demonstration should only have the logical meaning of a formal derivation, and he was uninterested in the psychological issue of evidence or the rhetorical conception of a proof. In fact, he strongly opposed the objections raised against the "excessive demonstrative rigor" of the *Elements*, in which one proves theorems (not to say the axioms themselves) which are intuitively known even by a child or a beast:

> I am very far from sharing the opinion of my friend Arnauld and other excellent men who, handing geometry over to the minds of students in a somehow relaxed and less severe way, are not content with Euclid's severity, as though it were useless, and criticize him too harshly. In doing so, they do not realize how important it is to achieve the truths not by the imagination (which even beasts are endowed with) but by the necessity of reason instead: and there is much that goes by between the education of students and a thoroughly complete analysis, in which nothing that can be demonstrated is assumed without demonstration.[6]

[6] The quotation is taken from the geometrical essay *Demonstratio omnimoda*, written in 1712 (in De Risi 2007, pp. 602–603). The text goes on with an answer to the old Epicurean criticism of Euclid, that in proving that a side of a triangle is shorter than the sum of the other two (*Elements* I, 20), he was attempting to prove something that every beast knows when it goes to food in a straight line (Proclus, *In Euclidis* 322–23; but the criticism was commonplace in Port-Royal epistemology, and the proposition was quoted as the paradigmatic example of the fault of "Prouver des chose qui n'ont pas besoin de preuves": cf. above *note* 3). Here is Leibniz' answer: "Et hinc Euclides bene demonstravit in triangulo duo latera semper esse tertio majora, et male objectum est à quibusdam id etiam bruta nosse, quae recta non per triangulum ad praesepe fluantur. Neque enim quaeritur quid exhibeat imaginatio (Divina Sapientia nobis animalibusque aliis multa per instinctum quemdam certè citra ratiocinationem insinuante, quae difficilius ratiocinando assequeremur) sed quid per necessarias consequentias ostendi posset. Si Euclides rectam definisset brevissimam inter duo puncta, non opus habuisset hac demonstratione pro lateribus trianguli: sed ille hac definitione usus non est, et fortasse si uti voluisset, difficilius inde alias rectae proprietates demonstrasset". Cf. similar Leibnizian statements ranging from a letter to Vagetius dating December 12th, 1679 (A ii, 1, n. 218, p. 769), related to Leibniz' work on geometry of the same months (LH XXXV, i, 2, Bl. 6–7; Echeverría 1980, vol. 2, p. 44), to the late *In Euclidis* πρῶτα from 1712 (GM v, p. 185). See also *Nouveaux Essais*, II, xxi, § 3: "C'est à peu près comme à l'egard des Axiomes vulgaires qui pourroient estre et qui meritoient d'estre demonstrés parmi les Theorêmes, et qu'on laisse passer cependant pour axioms comme si c'estoient des verités primitives. Cette indulgence nuit plus qu'on ne pense" (A vi, 6, n. 2, p. 170). As for Malebranche, Leibniz criticizes his attitude toward Savile

On the contrary, Leibniz thought with Clavius that the perfection of geometry demands that every axiom should be demonstrated and anything which is capable of proof effectively proved. Innumerable times he evoked the bright examples of Apollonius and Roberval, who had devoted a great deal of their efforts to improving the demonstrative rigor and proving the principles of geometry:

> Ramus blamed Euclid saying that he, looking for the rigor of proofs, had lost the most apt method to enlighten the mind; yet, the good Ramus, who wanted to change the Euclidean method, has lost both the rigor and the truth. ... The most common defect, for which not even Euclid is blameless, is to assume axioms that can be proved. It is true that this defect does not damage the certitude, if the axioms are justified by a multiplicity of experiences; and those of mathematics are such. This defect, however, does damage the perfection of the mind, and it is the main reason why we have not yet been able to transform the synthesis of geometers into an analysis. ... I am convinced that for the perfection of a science one should prove some of the propositions that are called axioms, as Apollonius did with some that Euclid had assumed without demonstration. Euclid was right, but Apollonius was more right. It is not necessary to do this, but it is important, and even necessary in some respects. The late Mr. Roberval had conceived some new *Elemens de Geometrie*, in which he rigorously proved many propositions that Euclid had assumed or taken for granted. I don't know whether he completed his work before he died, but I do know that many people mocked him. Had they known the importance of such work, they had judged otherwise. It is not necessary for the beginners, not even for the common teachers, but there is nothing more necessary than doing this for the advancement of science and to proceed beyond the Pillars of Hercules.[7]

---

in his reading note to the *Recherche de la vérité* (A vi, 4B, n. 348, p. 1847), adding that Euclid's definition of parallel lines is defective (p. 1848).

[7] The quote comes from the *Projet et essais pour avancer l'art d'inventer* (1688–1690), in A vi, 4A, n. 205, pp. 968–70. The reference to Apollonius comes from a remark of Proclus, *In Euclidis* 183, stating that the ancient geometer had attempted to prove several Euclidean axioms, and in particular the Common Notion that things equal to the same thing are equal to each other; to which Geminus had objected (in Port-Royal fashion) that Apollonius could not hope to establish by means of less well known middle terms what everyone knows. Proclus accepted Geminus' criticism and reported Apollonius' proof to show its futility in *In Euclidis* 194–95. Leibniz, of course, sided with Apollonius against Geminus and Proclus, and attempted himself to prove the same Common Notion (through a different demonstrative strategy). Among other passages in which Leibniz mentions Apollonius' or Roberval's endeavor, we may remember the above-mentioned letter to Vagetius from 1679 (in A ii, 1, n. 218, pp. 769 and 771); the *Demonstratio axiomatum Euclidis* from the same year: "Apollonium olim Axiomata demonstrare aggressum narrat Proclus, idem intelligo conatum facere Robervallium. Mihi recte fecisse videntur: tum demum enim ad perfectissimas comprehensiones perveniemus, cum nihil sensui aut imaginationi fidentes, omnia ad rationes exigemus" (A vi, 4A, n. 54, p. 165); the *Animadversiones* to Descartes' *Principia* (1691; to *Principia* i, § 1), in which Leibniz criticizes the French philosopher for not having attempted to prove the axioms: "Rem reique usum optime explanabimus exemplo Geometrarum. Constat esse apud eos Axiomata et Postulata, quorum veritati reliqua innituntur. Haec admittimus, tum quod statim satis faciunt animo, tum quod infinitis experimentis comprobantur, et tamen interesset ad perfectionem scientiae ut demonstrarentur. Hoc olim aggressi sunt in nonnullis Apollonius et Proclus, nuper Robervallius. Et sane quemadmodum demonstrare Euclides voluit Trianguli duo latera simul esse tertio majora (quod ut quidam veterum jocabatur, etiam asini norunt recta, non per ambages ad pabulum tendentes), quia scilicet volebat veritates Geometricas non imaginibus sensuum sed rationibus niti, ita poterat quoque demonstrare duas rectas (quae productae non coincidunt) unicum tantum punctum commune habere posse, si bonam rectae definitionem habuisset. Et magnum ego usum demonstrationis axiomatum esse scio ad veram analyticen seu artem inveniendi. Itaque si voluisset Cartesius id exequi, quod in praecepto ejus optimum est, debuisset elaborare in demonstrandis principiis sci-

In other papers, less radical than the one quoted above, Leibniz agreed that the proof of the axioms is not strictly necessary in geometry, and that the so-called evident principles may well be assumed *provisionally* by mathematicians (as their evidence make harmless their acceptance, dispelling the risk of a hidden contradiction). Thus, it is possible that geometry might leave its axioms unproven for many centuries. Still, they *can* and *should* be demonstrated for the perfection of science.[8]

---

entiarum, et agere in Philosophia, quod Proclus volebat in Geometria, ubi minus necesse est. Sed nostro autori potius visum alicubi est applausum quam certitudinem habere" (GP IV, pp. 354–55); a letter to Foucher from January 1692: "Quand j'estois à Paris, on se moquoit de M. Roberval, parce qu'il avoit voulu demonstrer quelques axiomes d'Euclide. Mais je ne m'en moquois pas; on sçait que Proclus, et même Apollonius y avoient déja pensé. Ceux qui aiment à pousser le detail des sciences, meprisent les recherches abstraites et generales; et ceux qui approfondissent les principes entrent rarement dans les particularités. Pour moy j'estime egalement l'un et l'autre, car j'ay trouvé que l'analyse des principes sert à pousser les inventions particulieres" (A II, 2, n. 137, pp. 490–91; a similar statement was already drafted in a letter addressed to Foucher in 1676 and never sent: see A II, 1, n. 120, p. 389); an important letter to Johann Bernoulli from the 2nd of September 1696: "Quod dixi omnis axiomatis a me demonstrationem desiderari, non temere dictum est; idque animadvertes opinor, si quando vacabit inspicere meditationes quasdam meas de ideis, quae extant in Lipsiensium actis. Excipio tamen axiomata illa quae sunt indemonstrabilia ipsas scilicet identicas propositiones. Caetera omnia, quae scilicet possunt demonstrari, etiam utile est demonstrari, cum aliqua magni momenti theoremata in iis fundantur idque etiam veteres viderunt. Unde Apollonius (in scriptis deperditis), et Proclus et alii axiomata ab Euclide assumta demonstrare sunt conati. Eamque rem fructu non carere facile opinor concedes quem tamen non vident, qui scientiarum utilitatem vulgari modulo metiuntur. Interim vides, ea limitatione quam addidi et quam addendam esse praevideri poterat, non esse cur progressum in infinitum vereare in demonstrando" (A III, 7, n. 29, p. 113); the *Nouveaux Essais*, IV, VII, § 1: "Vous trouverez en cent lieux que les Philosophes de l'Ecole ont dit que ces propositions sont evidentes *ex terminis* aussi-tôt qu'on en entend les termes; de sorte qu'ils étoient persuades que la force de la conviction étoit fondée dans l'intelligence des termes, c'est à dire, dans la liaison de leurs idées. Mais les Geometres ont fait bien d'avantage: c'est qu'ils ont entrepris de les demontrer bien souvent. Proclus attribute déja à Thales de Milet, un des plus anciens Geometres connus, d'avoir voulu demontrer des propositions qu'Euclide a supposes depuis comme evidentes. On rapporte qu'Apollonius a demontré d'autres Axiomes, et Proclus le fait aussi. Feu M. Roberval, déja octuagenaire, ou environ, avoit dessein de publier de nouveaux *Elemens de Geometrie* don't je croi vous avoir déja parlé. Peut etre que les *Nouveaux Elemens* de M. Arnaud, qui faisoient du bruit alors, y avoient contribué" (A VI, 6, n. 2, p. 407; cf. a similar passage in I, III, § 24, p. 107); as well as the *Recommandation* (see below, *note* 8).

[8] Leibniz offered the best survey of his insights on the need to prove the Euclidean axioms (and his anti-Cartesianism on the topic) in a paper from 1686 called *Recommandation pour instituer la science generale*, which is worth quoting at length: "On voit même que Proclus et autres Geometres tachent de donner des demonstrations de quelques Axiomes, dont personne ne doute, et qu'Euclide a crû de pouvoir supposer, comme par exemple, que deux droites ne sçauroient avoir un segment commun. C'estoit aussi l'opinion de feu M. de Roberval, qu'il falloit demonstrer les Axiomes mêmes autant qu'on peut, ce qu'il vouloit faire effectivement à ce que j'ay ouy dire dans les *Elemens de Geometrie* qu'il avoit projettés. Et chez moy ce soin de demonstrer les Axiomes est un des plus importans points de l'art d'inventer, dont je diray les raisons une autre fois, me contentant maintenant d'en faire mention, à fin qu'on ne s'imagine point que ce travail seroit inutile et ridicule; et parce que c'est en effect un corollaire du grand precepte que je viens de donner. Et c'est un de mes estonnemens de voir que ce philosophe celebre de nostre temps, qui a tant recommandé l'art de douter, a si peu mis en usage ce qu'elle contient de bon, dans les occasions, où il auroit esté le plus utile: se contentant d'alleguer l'evidence pretendue des idées; à quoy Euclide et les autres Geometres ont fort sagement fait de ne se pas arrester; aussi est ce le moyen de couvrir toutes sortes

Leibniz added that the practical need for axioms is mostly due to the imperfection of the definitions, which sometimes are not able to spell out clearly all the features of the defined object. Thus, for instance, Euclid's infamous definition of a straight line as the line that "lies evenly with the points on itself", is almost useless as it gives no real clue about the geometrical properties of a line. Geometers are compelled to employ axioms to express some of the properties of such a line, such as that two straight lines do not encompass any space, or that a straight line can be extended in both directions. Similarly, a defective definition of parallels (like the one employed by Euclid) may force a mathematician to accept the Parallel Postulate. Given an unaccomplished system of definitions, thus, axioms are necessary for the construction of a rigorous proof, as they express a few properties of the involved objects. But, Leibniz thought, a fully developed set of definitions should be able to get along without the need of any axiom, and in particular the actual axioms employed in the various editions of the *Elements* should be proved starting from good definitions. The defective definitions, moreover, are also the principal cause of the lack of rigor in several classical geometrical proofs. The axioms normally employed, in fact, do not capture the entire essence of the terms involved, and the geometer has to rely on representations and images. Diagrammatic inferences, unlicensed by an explicit set of principles of geometry, are inserted in the proofs since the imagination has to supply what the conceptual definitions had left undetermined. Needless to say, this may be especially damaging when dealing with the Parallel Postulate, which is a proposition quite evident to the imagination and yet difficult to prove in logical form.[9]

de visions et de prejugés. Cependant j'accorde qu'on peut et qu'on doit souvent se contenter de quelques suppositions, au moins en attendant qu'on en puisse faire aussi des theoremes un jour, parce qu'autrement on s'arresteroit trop quelque fois. Car il faut tousjours tacher d'avancer nos connoissances, et quand même ce ne seroit qu'en establissant beaucoup de choses sur quelque peu de suppositions, cela ne laisseroit pas d'estre fort utile. Car au moins nous sçaurions, qu'il ne nous reste à prouver que ce peu de suppositions pour parvenir à une pleine demonstration, et en attendant nous en aurons au moins d'hypothetiques, et nous sortirons de la confusion des disputes. C'est la Methode des Geometres, par exemple Archimede suppose que la droite est la plus courte des lignes, et que de deux lignes d'un même plan, dont chacune est partout cave d'un même costé, l'incluse est moindre que l'includente, et là dessus il acheve rigoureusement ses demonstrations. Mais il est fort important de faire expressement toutes les suppositions dont on a besoin, sans se donner la liberté de les prendre tacitement pour accordées, sous pretexte que la chose est evidente d'elle même par l'inspection de la figure, ou par la contemplation de l'idée. A quoy je trouve qu'Euclide, tout exact qu'il est, a manqué quelques fois, et quoyque Clavius y ait souvent suppléé par sa diligence, il y a des endroits, où il n'y a pas pris garde, dont un des plus remarquables et des moins remarqués se rencontre d'abord dans la demonstration de la premiere proposition du premier livre, où il suppose tacitement que les deux cercles qui servent à la construction d'un triangle equilatère, se doivent rencontrer quelque part, quoyqu'on sçache que quelques cercles ne se sçauroient jamais rencontrer. Mais on ne se trompe pas aisement en Geometrie par ces sortes de suppositions tacites. Les Geometres ont trop de moyens de decouvrir les moindres erreurs, si par mégarde ils leur en echappoient" (A VI, 4A, n. 161, pp. 703–705).

[9] On the fact that Euclid was compelled to assume the Parallel Postulate because he had a defective definition of parallel lines, see the *In Euclidis* πρῶτα (GM V, pp. 201–202; *text 36* below). On the connection of this issue with the imagination and the use of diagrams, see *Nouveaux Essais*, IV, XII, § 4: "Euclide, par example, a mis parmi les Axiomes ce qui revient à dire, que deux lignes droites ne se peuvent rencontrer qu'une seule fois. L'imagination prise de l'experience des sens,

Leibniz in fact went further than any of his predecessors in disregarding imagination in geometry, and he raised the standards of rigor in this science quite above those normally accepted at the time. Not only did he warn about the use and abuse of diagrams, that may suggest wrong deductions or induce a relaxed demonstrative attitude; ideally, he would have liked to dispense with them entirely. Leibniz wanted to present all geometrical proofs (including those of axioms) as logical arguments *in forma* or chains of syllogisms (or other logical inferences) starting from the definitions. This kind of reduction of Euclid to syllogistic reasoning had already been attempted, it is true, in the past. The celebrated *Analyseis Euclideae* (1566) by the mathematician Christian Herlinus and his pupil Konrad Dasypodius had displayed in fact the first six books of the *Elements* as chains of syllogisms (or other propositional inference rules); and Clavius himself had quoted with praise the *Analyseis* in his commentary, reproducing Herlinus' logical proof of *Elements* I, 1 and stating that a similar presentation of the subject could be attained for the whole of mathematics.[10]

ne nous permet pas de nous figurer plus d'une rencontre de deux droites, mais ce n'est pas sur quoi la science doit être fondée. Et si quelqu'un croit que cette imagination donne la liaison des idées distinctes, il n'est pas assez instruit de la source des verités, et quantité de propositions demontrables par d'autres anterieurs passeroient chez lui pour immediates. C'est ce que bien des gens qui ont repris Euclide, n'ont pas assez consideré: ces sortes d'images ne sont qu'idées confuses, et celui qui ne connoit la ligne droite que par ce moyen, ne sera pas capable d'en rien demontrer. C'est pourquoi Euclide faute d'une idée distinctement exprimée, c'est à dire d'une definition de la ligne droite (car elle qu'il donne en attendant est obscure, et ne lui sert point dans ses demonstrations) a été obligé de recourir à deux Axiomes qui lui ont tenu lieu de definition et qu'il employe dans ses demonstrations, l'un que deux droites n'ont point de partie commune, l'autre qu'elles ne comprennent point d'espace. … Il y a aussi de l'apparence qu'en suivant les sens et leurs images, on seroit tombé dans des erreurs; à peu prés comme l'on voit que tous ceux qui ne sont point instruits dans la Geometrie exacte, reçoivent pour une verité indubitable sur la foi de leur imagination que deux lignes qui s'approchent continuellement, doivent se rencontrer enfin. Au lieu que les Geometres donnent des instances contraires dans certaines lignes qu'ils appellent Asymptotes. … Euclide a eté obligé pourtant de se borner à certaines Axiomes, dont on ne voit pas l'evidence que confusement par le moyen des images" (A vi, 6, n. 2, pp. 451–52). Both axioms quoted by Leibniz are in fact spurious and are not included among the original Euclidean postulates. The last "truth", indubitable for the imagination, that two converging (straight) lines will meet, is of course another form of the Parallel Postulate, and was taken as a substitute axiom for it by Arnauld in the *Nouveaux elemens* (see above, *note* 3).

[10] See Clavius, *Euclidis*, pp. 28–29, that followed (with a few changes) Herlinus, *Analyseis*, p. 2v. This was a common opinion and Barrow, for instance, maintained the same: "Saltem haud inutile est, nec injucundum, ipsarum inter se primarum propositionum, operâ sillogisticâ cohaerentiam et cognationem ostendere" (*Lectiones mathematicae*, iv, p. 71; but see his later discussion about the principles, the evidence of which is witnessed "propriae conscientiae testimonio certo", in *Lectiones*, v, p. 78; while in *Lectiones*, vii, p. 115, Barrow explicitly said that axioms can be proved from definitions and they are different from theorems since their proof is *immediate*; and for this very reason, he also stated in *Lectiones*, viii, p. 128, that the Parallel Postulate is in fact a theorem rather than an axiom). Jungius' disciple Johannes Vagetius published an emended version of Clavius' syllogistic proof of *Elements* I, 1 in his *Admonitio* to his master's *Logica* (see a modern edition of this text in Risse 1977, pp. 257–61). The latter proof had a great relevance in the history of logic, since Vagetius claimed that mathematical theorems can be demonstrated in a purely logical way, but not by syllogisms only: one has to extend the number of valid logical inferences beyond those recognized by Aristotle; Leibniz agreed with him (see Mugnai 2010). Later on, the reducibility

The logical regimentation of the *Analyseis*, however, only applied to the propositional part of the Euclidean proof, leaving a large amount of informal deduction to the inspection of the diagrams (to "diagrammatic reasoning"); and only in this way could Dasypodius and his master claim to have attained a logical and gapless proof of so many theorems of elementary geometry (they did not try to prove the axioms).[11] Other authors, like Borelli and Wallis, had remarked that Euclid had to rely on diagrammatic inferences ("*ex inspectu Schematis*")[12] to supply principles that were not explicit enough in his axiomatization; instead of syllogizing Euclid like Clavius, Borelli had tried to provide a new, explicit set of principles that can be employed to fashion a gapless proof of *Elements* I, 1.[13] Leibniz intended to go further and sur-

of mathematical proof to syllogisms was one of the cornerstones of Wolff's mathematical epistemology, and was consequently criticized by Lambert and Kant. It should be noted, however, that Jungius himself was among the first logicians to acknowledge the existence of those "*consequences asyllogistiques bonnes*" that Leibniz will mention in the *Nouveaux Essais* (see below, *note* 16) and begin to study in his logical writings.

[11] In particular, Herlinus and Dasypodius accepted Grynaeus' Greek text, including the Parallel Postulate as Axiom 11, and a definition of parallels as non-incident straight lines, which they called however *aequidistantes* (*Analyseis*, pp. 1v–2r). They also accepted, in fact, Posidonius' definition of parallels as equidistant lines (in the proper sense) in their commentary to the definition (p. 90r). The *Analyseis* were known by Leibniz, who quoted them (for instance) in the important *Meditationes de cognitione, veritate et ideis* published in 1684: "De caetero non contemnenda veritatis enuntiationum criteria sunt regulae communis Logicae, quibus et Geometrae utuntur, ut scilicet nihil admittatur pro certo, nisi accurata experientia vel firma demonstratione probatum; firma autem demonstratio est, quae praescriptam a Logica formam servat, non quasi semper ordinatis scholarum more Syllogismis opus sit (quales Christianus Herlinus et Conradus Dasypodius in *sex priores Euclidis libros* exhibuerunt) sed ita saltem ut argumentatio concludat vi formae, qualis argumentationis in forma debita conceptae exemplum, etiam calculum aliquem legitimum esse dixeris" (A VI, 4A, n. 141, pp. 590–91). Note, however, that in a fragment from 1679 Leibniz referred to the syllogistic rendering of the *Elements* as the work of Clavius, probably still ignoring Herlinus' book (*Demonstrationes geometrarum non satis severas esse*, LH XXXV, I, 4, Bl. 1–2; in ECHEVERRÍA 1980, vol. 2, pp. 309–10). In this last remark, moreover, Leibniz was unsatisfied with the formal proof of *Elements* I, 1, as Herlinus and Clavius had not detected any problem in the intersection of the two circles, plainly relying on diagrammatic inferences (see the following *note* 13). In *Nouveaux Essais*, IV, I, § 9 (A VI, 6, n. 2, p. 361), however, Leibniz seemed to praise Herlinus' work just because it went a step forward toward a purely conceptual approach to geometry. Another source for Leibniz was probably Vagetius' proof mentioned in the previous *note*, that was published in 1681.

[12] The expression and quote comes from Wallis' *De postulato quinto* (p. 678).

[13] The construction of an equilateral triangle given in *Elements* I, 1 assumes (among other things) that two circles intersect each other in two points: but apparently Euclid had to prove that the circles are in such a relative position to overlap one another, that (in such a position) they actually meet thanks to some principle of continuity, and that they meet in exactly two points (and not in a segment, for instance). In all these passages, Euclid (and Herlinus and Clavius after him) seemed to rely on implicit assumptions which are evident from the inspection of the diagram. Since Antiquity, however, these implicit assumptions had been object to criticism (see PROCLUS, *In Euclidis*, 214–18), and *Elements* I, 1 remained for several centuries the paradigmatic proposition to discuss foundational problems about the logical cogency of the proofs in elementary geometry. In the seventeenth century, the best foundational efforts in this direction were made by Giovanni Borelli, who in the first edition of his *Euclides restitutus* (1658) explicitly assumed an Axiom 13 to guarantee continuity enough to have the two circles intersect, although he relied on the diagram to ascertain that the two circles are in such a position to overlap. In the 1679 edition of the same work, Borelli

pass all these previous attempts, renouncing any help from the imagination and the diagram, and proving all the extra-principles that could be envisaged to adjust the standard proofs; he wanted to produce a purely intellectual science.[14]

To this effect, he devised a new formalism for geometry (the *characteristica geometrica* or *characteristica situs*) that allowed him to dispense with inspecting the diagram in a demonstration.[15] Most of all, he was looking for a new logic that

---

gave a slightly more complex construction to explicitly prove that the two circles overlap as for their position, but now he thought that he did not need anymore an additional continuity principle (he renounces his Axiom 13). A few years before, Pascal had given a few principles to this effect in his *Introduction à la géométrie* (*Oeuvres*, vol. 9, pp. 292–94), which remained unpublished but was read by Leibniz. Caspar Schott had added an Axiom 26 to the *Elements* to guarantee that the two circles meet. In the following century, Abraham Kästner assumed a similar Axiom 7 in his *Anfangsgründe der Arithmetik,* p. 189. See also De Risi [c]. The entire discussion is very relevant from a foundational point of view, as such principles seem to be the first instances of the so-called Circle-Circle continuity axiom (which is equivalent to Line-Circle continuity). This axiom, supplemented with the Archimedean Axiom or at least the Principle of Aristotle (see below for Leibniz' treatment of both) gives in fact continuity enough to recover most of the classical Euclidean geometry (cf. Hartshorne 2000 and Greenberg 1979, 2010). Leibniz, besides transcribing Pascal's work, and commenting on Schott's book (in Echeverría 1995, p. 60; cf. *text 6*), underlined the whole discussion in his copy of Borelli's book, and discussed the issue quite often; see for instance his *Demonstratio omnimoda* (1712) in De Risi 2007, p. 602, where he referred to Borelli's reform of *Elements* I, 1; or a similar passage in the *Specimen geometriae luciferae*, GM vii, p. 284.

[14] For an explicit quote, see the *Nouveaux Essais* in *note* 9, to which one may add several other passages of the same work (e. g. II, xxix, §§ 13–16, in A vi, 6, n. 2, pp. 261–63). Among Leibniz' many other declarations about geometry as a purely intellectual science that should not rely on diagrams, see for instance: "Ces sont ordinairement des gens, qui se sont plus tost servi de leur imagination que de leur raison; et qui n'ont jamais rien compris par demonstration, mais seulement par experience, ou opinion. C'est pourquoy ils ne sauroient comprendre que d'autres puissent estre plus asseurez qu'eux. Mais il n'y a point d'autre moyen de les guerir, que de les renvoyer à Euclide ou Archimede, à fin d'apprendre que la source de la certitude de la geometrie n'est pas dans les figures, mais dans les idées abstraites des choses incorporelles: et que par consequent il y a de la certitude dans des matieres même où les figures n'ont pas lieu" (to the Duke of Hannover, beginning of 1677; A ii, 1, n. 134, p. 470).

[15] Truth be told, Leibniz' position about the use of imagination in geometry was far more complex and nuanced than the one I have briefly sketched. First of all, he thought that imagination plays an important role in discovering mathematical truths (the *ars inveniendi*), even if it should not be employed in the context of justification. Second, in his metaphysical writings Leibniz considered space itself (the very object of geometry) as a product of imagination. Leibniz' late monadological idealism, thus, seems to imply that geometry is in fact ontologically grounded in the power of imagination. Third, Leibniz' epistemological writings also verged in the same direction, as he was well aware that any *characteristica* or form of symbolic reasoning needs an imaginative realization of the symbols employed, and the combinatorial rules that govern the syntax of the formal language have to be implemented in a sensible *medium*. The conceptual tool that Leibniz envisaged to bridge the gap between his logical understanding of mathematics and his metaphysical and epistemological use of imagination, is that of an *imaginatio distincta* that has hybrid properties between imagination proper and the intellect. The term surfaced already in the *De ortu, progressu et natura algebrae* from 1685 (Gm vii, pp. 203–16), and was then widely used in the correspondence with Bodenhausen from 1690–1698 (originally published in GM vii, pp. 353–93 and now to be found in A iii, 4–7). An important text on the connection between *analysis situs* and imagination is the *Ars representatoria* from 1691, to be found in De Risi 2007 (pp. 580–81) and (with an English translation) in De Risi 2005.

surpassed the strictures of classical syllogistic (or basic propositional calculus) and was able to open toward a logic of relations.[16] With the help of these tools, he was able to attempt several proofs of the Euclidean Common Notions, i.e. those more abstract axioms that deal with the concept of quantity in general and cannot be considered geometrical in the proper sense. Leibniz' favorite examples are his proofs that two magnitudes equal to a third are equal to one another, and that the whole is bigger than the part, but he was able to devise many more of them.[17] Coming to the proper geometrical theorems and axioms (including the Parallel Postulate), however, his project began to run into innumerable difficulties, as his *characteristica* remained at an embryonic stage and his logic was still quite underdeveloped. In all these cases he accepted (provisionally or not) more tentative proofs and ended up by relaxing his high standards of rigor. We will see, however, how Leibniz tried sometimes purely symbolic proofs even for the Parallel Postulate, and how, in any case, he could not content himself with Clavius' proofs of the postulates, which in fact relied on vague notions (such as the above-mentioned "uniformity") and an unscrupulous use of intuitive constructions, diagrammatic reasoning and kinematic arguments.

3.2 Leibniz' Theory of Definition.

Leibniz' logicist project in the foundations of geometry was grounded (as he said) on his definition of truth as containment of the notion of the predicate in that of the subject of a proposition; that is, on an *analytical* conception of truth. The aim of a proof of a statement (take the judgement 'a square has four sides') is thus to show that the

[16] A famous passage in *Nouveaux Essais*, IV, XVII, § 4, witnessed to posterity Leibniz' opinion about the strictures of syllogistics: "De plus il faut scavoir qu'il y a des *consequences asyllogistiques bonnes* et qu'on ne scauroit demonstrer à la rigueur par aucun syllogisme sans en changer un peu les termes; et ce changement même des termes fait la consequence asyllogistique" (A VI, 6, n. 2, p. 479).

[17] Leibniz gave several proofs of the Euclidean Common Notions (the ones mentioned above are the first and the eighth in the *Elements*, the latter being probably spurious), and the theme is recurrent in his writings. We have already seen a remark on this subject in the *note* 7. An early writing mostly devoted to the issue is the *Demonstratio axiomatum Euclidis* from 1679 (A VI, 4A, n. 54, p. 165–67), but the topic is treated again and again in the following years. Among the unpublished papers, the *Expediendi laboris causa* (a section of which we are editing as *text 25*) is one of the longest and most relevant essays devoted to the proof of the Common Notions. We may also mention a programmatic statement written in the margins of Leibniz' copy of Barrow's edition of Euclid, stating: "Aequalia sunt quae magnitudine discerni non possunt, seu quae sibi substitui possunt magnitudine salva. Ex hoc solo axiomata omnia demonstrantur" (in BARROW, *Euclidis*, p. 8). Barrow himself had attempted a similar task in his *Lectiones*, p. 203. On Leibniz' attempts in this direction, see FICHANT 1992. Since several mathematicians attempted to prove all the geometrical principles, Leibniz was not alone in his attempts to prove (for instance) that the whole is bigger than its part. See at least Clavius' most naïve treatment in *Euclidis*, p. 24; and Saccheri's opinion that it should be proved in order to be accepted in science: "Omnia axiomata (exceptis duobus universalissimis) famulari debent definitioni *quid nominis*, & ex ea confirmari. Sit axioma *totum est maius sua parte*. Dico admittendum non esse, nisi post definitiones *quid nominis* terminorum, ex quibus demonstrativè elici possit" (*Logica demonstrativa*, p. 201).

concept of the predicate ('four-sided') is identical with a conceptual mark contained in the concept of the subject ('square'). The deduction itself is thus performed (in its most explicit form, which is not always needed in practice) through an analysis of the subject and the predicate into their conceptual constituents, which is effected through a system of definitions (a definition of the subject, then definitions of the terms involved in the definition of the subject, and so forth; in our example: 'def: a square is a quadrilateral with equal sides', and 'def: a quadrilateral is a four-sided figure'); and a relation of containment which is reduced to the principle of identity of the conceptual marks found in the analysis of the subject and the predicate ('a four-sided figure with equal sides is four-sided'). The entire proof is thus ultimately based on the original definitions of the geometrical terms and law of identity ('four-sided is four-sided').

Leibniz thought that the conceptual analysis of the notions involved in a statement may eventually arrive at a set of concepts that cannot be further analyzed (the *simple ideas*). To guarantee the truth of a proposition, however, it is not necessary to reach the bottom of analysis and resolve every concept into simple ideas (a task that may prove exceedingly difficult or even impossible), and it is enough to analyze both the subject and the predicate of a judgment to the point at which the predicate is found to be identical to an (intensional) part of the subject; that is, the decomposition of the notions involved takes the form 'AB is A'; a proposition in this form, on the other hand, is analytically true thanks to the true identity 'A=A'.[18] The law of identity is expressed by Leibniz as a set of identical axioms which should be considered, from the strictest foundational point of view, as authentic and indemonstrable principles:

> All propositions, however, whose truth must be shown by further analyzing and understanding their terms are demonstrable by such analysis, that is, by definitions. So it is clear that demonstration is a chain of definitions. For in the demonstration of any proposition, nothing

[18] Among Leibniz' statements about the difficulty of a complete analysis of notions, see for instance: "Les premiers Terms indefinibles ne se peuvent, aisement reconnoistre de nous, que comme les nombres premiers: qu'on ne sçauroient bien reconnoistre que negativement et comme par provision" (*Sur les premieres propositions*, 1676; in A vi, 3, n. 53, p. 436). Complete analysis is not necessary: "Il est très difficile de venir à bout de l'analyse des choses, mais il n'est pas si difficile d'achever l'analyse des verités dont on a besoin. Parce que l'analyse d'une verité est achevée quand on en a trouvé la demonstration: et il n'est pas toujours necessaire d'achever l'analyse du sujet ou du prédicat pour trouver la demonstration de la proposition" (*De la sagesse*, 1676; in A vi, 3, n. 89_6, p. 671). See also his reference to Pascal in the *Paraenesis de scientia generali tradenda*: "De discrimine inter conceptus imperfectos et perfectos, ubi occurritur difficultati Pascalii de Resolutione continuata et ostenditur ad perfectas demonstrationes Veritatum non requiri perfectos conceptus rerum" (A vi, 4A, n. 206, p. 974). It is even possible that Leibniz may have regarded some ideas as simple in a given context, and still analyzable in other fields of enquiry. It is true, however, that even though a proposition may be proved analytically true without resolving its terms into simple ideas, a complex idea may still hide a contradiction (a term may be ill-defined); and the apparently true proposition 'AB is A' may still be false since A (for instance) may be an inconsistent notion (its further analysis would disclose it to be of the form 'C & non-C'). Thus, an analysis of the terms of a proposition that reaches the canonical form 'AB is A' may ascertain the truth of it: either if it is complete and A and B are simple ideas, or if one can claim the consistency of the complex notions A and B by other means (e.g., an object falling under A exists and it is empirically observed). See for instance a deleted passage from the *Meditationes de cognitione, veritate et ideis*, in A vi, 4A, n. 141, pp. 590. On the Leibnizian analysis of notions in relation to mathematics, see Rabouin 2015.

> is used but definitions, axioms (with which I here include postulates), theorems which have been demonstrated previously, and observations. Since the theorems again must themselves be demonstrated, it follows that *all truths can be resolved into definitions, identical propositions, and observations* – though purely intelligible truths do not need observations. After the analysis has been completed, it will become manifest that the chain of demonstration begins with identical propositions or observations and ends in a conclusion but that the beginning is connected with the conclusion through intervening definitions.[19]

This is the general logical framework that Leibniz employed to justify his mathematical epistemology and his program of proving all geometrical proposition without having recourse to diagrams or other non-logical inferences. This analytical procedure of proof should be applied to all the Euclidean theorems, and to all the Euclidean axioms and postulates. Leibniz thus claimed that his project in mathematics was not so much that of proving all the axioms as that of reducing all the composite (and commonly used) axioms to the simple, identical ones.[20] This acknowledged, it is still

[19] From a letter to Conring, March 29th, 1678 (A II, 1, n. 168, p. 602; transl. LOEMKER 1956). In the same letter, Leibniz writes about the fact that *we don't have* to prove the usual axioms of geometry, as they are evident enough (even though they can be proved), but *we cannot* prove the identical axioms: "Axiomata mihi videntur propositiones quae vel non possunt vel quas non necesse est demonstrari. Demonstrari non possunt illae, quae sunt identicae, verbi gratia: *A est A*, *unumquodque sibi ipsi aequale est*, et similes infinitae (tot enim dantur quot termini) quas qui negat eo ipso frustra ratiocinari sese ac disputare fatetur, sublata enim erit inquisitio veritatis si idem verum et falsum esse possit. Demonstrari necesse non est alias quae ab auditoribus omnibus sine difficultate admittuntur, ut *totum esse majus parte*; quae etsi formaliter identicae non sint, facile tamen ad identicas reduci id est demonstrari possunt" (p. 599). Leibniz had already communicated to Conring his ideas on the topic on January 13th: "Axiomata ego non ut ais ἀναπόδεικτα sed tamen plerumque non necessaria demonstratu esse arbitror. Demonstrabilia vero esse pro certo habeo. Unde enim constat nobis de eorum veritate? non, opinor ex inductione, ita enim omnes scientiae redderentur empiricae, ergo ex ipsismet, id est ex earum terminis: quod fit vel quando idem dicitur de seipso (ex. causa, *A est A*, *unumquodque sibi ipsi aequale est*, et similes identicae) vel quando sola terminorum significatione sive quod idem est definitione, intellecta statim apparet propositionis veritas. Omnes ergo propositiones certae demonstrari possunt praeter identicas et empiricas" (A II, 1, n. 162, p. 582). We have already mentioned the letter to Vagetius from the following year: "Mea sententia solae propositiones identicae sunt indemonstrabiles et vere primae, caeterae omnes necessariae ab iis pendent. Axiomata Euclidis (si aequalibus addas aequalia, etc.), pendent ex definitione aequalium, id est eorum quorum magnitudo est [eadem]. Eadem, inquam, id est, ut sibi substitui possint magnitudine salva" (A II, 1, n. 218, p. 769). In any case, Leibniz' theory of demonstration from definitions and identical propositions is very recurrent in his logical writings. See at least the fragment *De demonstratione axiomatum non identicorum*, from 1682–1684: "Demonstrari possunt, non tantum theoremata, sed etiam ipsa Axiomata quae identica non sunt ..." (A VI, 4A, n. 119, p. 506); then § 61 of the *Generales inquisitiones de analysi notionum et veritatum* (A VI, 4A, n. 165, p. 759); and finally a long, famous passage in *Nouveaux Essais*, IV, VII (A VI, 6, n. 2, p. 406–27), which also presents a logical proof of 2+2=4 that was much discussed from the eighteenth century onwards.

[20] The statement 'A=A' is not properly an axiom, but an axiom schema involving as proper axioms all the identities of the defined terms. Thus Leibniz' ideal geometrical system has (at least) a countable number of identical axioms: "Il est tres important de concevoir que le nombre des premieres propositions est infini, car ells sont ou definitions ou Axiomes. Le Nombre des definitions aussi bien que des termes est infini. Le nombre des Axiomes l'est aussi. J'appelle *Axiome* proposition necessaire indemonstrable. *Necessaire* c'est à dire dont le contraire implique contradiction. Or la seule proposition don't le contraire implique contradiction, sans qu'on la puisse demonstrer, est

true that the entire burden of the proof is in fact resting on the definitions (since the proper axioms are just identities). We have noted that this was a common feature of seventeenth-century epistemology.

Leibniz, however, was much more sensitive than most of his contemporaries to delicate foundational issues, and was less inclined to be beguiled by axioms disguised as definitions. The latter should be stipulations about the use of words, and not assumptions about the reality of the defined objects. This is especially true in geometry and the mathematical sciences: Leibniz was able to clearly distinguish (like Borelli and others before him) a mathematical definition from the statement that the object defined "exists" (at least as an ideal entity). Thus, for instance, equidistant straight lines may be defined, but proving that there are such lines is another matter altogether (that may require a proof equivalent to the Parallel Postulate).

Since everything in mathematics (including the standard axioms) depends on definitions, and definitions are basically arbitrary stipulations, it may now seem that geometry itself is an entirely arbitrary construction.[21] Leibniz' classical answer to this query is that he is not a nominalist in a strong sense. Hobbes (he claimed) took mathematics to be a matter of conventions, since he saw it (correctly, Leibniz thought) as depending on definitions alone, and maintained (moreover) that definitions are just names given be arbitrary collections of conceptual marks. Leibniz wanted to correct this form of Hobbesian "super-nominalism", as he called it, by requiring that (proper) definitions be not just *arbitrary* collections of notions, but collections of notions that are not internally inconsistent.[22] The main point here is that Leibniz thought that the consistency of a definition is sufficient to guarantee the possibility and thus the (ideal) existence of the defined mathematical object. His

---

l'identique formelle. … Donc les Axiomes veritables et indemonstrables sont les propositions identiques. Or leur nombre est infini" (*Sur les premieres propositions*, 1676; in A vi, 3, n. 53, p. 435). Since these axioms cannot be expressed in purely logical terms (as they state the identity of *mathematical* notions), it may be questioned whether Leibniz' epistemological program in geometry is *logicist* in the strictest sense. It should be remarked, however, that the only *simple* mathematical concept involved in Leibniz' *analysis situs* should be (in the end) that of situation; the identical axioms express identities among situational structures. The concept of *situs* itself, on the other hand, may be considered (in certain respects) a purely logical concept.

[21] Here is an explicit quote on the dependence of the whole of geometry on arbitrary definitions: "Definitio enim ab arbitrio humano est. At vero omnia axiomata quae a sensu non pendent, imo omnia scientiarum a sensu et experimentis independentium theoremata sunt propositiones ejusmodi, quod et Aristoteles animadvertit, qui unicum posuit demonstrandi principium: definitionem. Et vero Axiomata omnia quae Euclides velut principia praemisit Elementis ex definitionibus demonstrabilia sunt" (Leibniz to Gallois, draft from the end of 1672; A ii, 1, n. 109, p. 351; A iii, 1, n. 2, p. 14). Leibniz is here discussing Hobbes' opinions.

[22] The label of super-nominalist (*plusquam Nominalis*) is attributed to Hobbes in the *Preface to Nizolius* (A vi, 2, n. 54, p. 428), but the criticism is recurrent. See for instance *De synthesi et analysi universali*: "Hobbesius scilicet cum videret omnes veritates posse demonstrari ex definitionibus, autem omnes definitiones esse arbitrarias et nominales crederet, quia in arbitrio est nomina rebus imponere, volebat et veritates in nominibus consistere, et arbitrarias esse. Sed sciendum est, non posse pro arbitrio conjungi notiones, sed debere ex iis conceptum formari possibilem, ut habeatur definitio realis; unde patet omnem realem definitionem continere affirmationem aliquam saltem possibilitatis" (A vi, 4a, n. 129, p. 542).

ontology of ideal objects, in fact, equated mathematical existence with possibility: if a geometrical definition is consistent, its object is possible and in fact it exists as such (as an ideal objectivity, an essence). This is not true, of course, outside the mathematical domain, where existence and possibility are no longer co-extensive notions. The notion of possibility as the existence of the essence of a thing, is usually called the *reality* (*realitas*) of the thing itself.

Thus, Leibniz distinguished "nominal definitions", which are arbitrary Hobbesian definitions, from "real definitions", which are also quite arbitrary (and thus Leibniz would endorse a mild form of nominalism), but show the *possibility* (i. e. the *realitas*) of their object.[23] The possibility of the object is simply the non-contradictoriness of the conceptual marks that are put together into the definition of the object itself.

Thus, real definitions should provide a proof of internal consistency. Moreover, as the analysis of the notions involved in the definition is often too complex to be completely carried out, Leibniz claimed that the simplest way to show the possibility of the defined object is to exhibit it *in concreto* (if it exists, it is also possible). The possibility of an empirical object may be ascertained through its actual existence (perceiving it, for instance). The possibility of a mathematical object, however, may be warranted through its *construction*. Ruler and compass constructions in classical Euclidean geometry, for instance, exhibit the reality of the corresponding mathematical objects, and thus count as proofs of the consistency of their definitions. A third and most perfect way to establish the consistency of a definition, besides empirical check

[23] Even though Leibniz envisaged the distinction between nominal and real definitions in his critical engagement with Hobbes, the most famous treatment of the topic is the one he gave against Locke in *Nouveaux Essais*, III, III, §§ 15–19: "L'essence dans le fonds n'est autre chose que la possibilité de ce qu'on propose. Ce qu'on suppose possible est exprimé par la definition, mais cette definition n'est que nominale quand elle n'exprime point en même temps la possibilité, car alors on peut douter si cette definition exprime quelque chose de reel, c'est à dire de possible; jusqu'à ce que l'experience vienne à nostre secours pour nous faire connoistre cette realité *a posteriori* lorsque la chose se trouve effectivement dans le monde, ce qui suffit au defaut de la raison qui feroit connoistre la realité *a priori* en exposant la cause ou la generation possible de la chose definie. Il ne depende donc pas de nous de joindre les idées comme bon nous semble, à moins que cette combinaison ne soit justifiée ou par la raison qui la monstre possible, ou par l'experience qui la monstre actuelle, et par consequent possible aussi" (A VI, 6, n. 2, pp. 293–94). He then continued (*Nouveaux Essais*, III, VI, § 28) advancing the instance of the impossibility of a regular decahedron: "il n'est pas en notre pouvoir de faire des combinaisons à notre fantasie, autrement on auroit droit de parler de *Decaedres reguliers* …" (A VI, 6, n. 2, p. 321). Another important geometrical example comes from § 24 of the *Discours de métaphysique*: "Il est bon aussi de discerner les definitions nominales et reelles, et j'appelle definition nominale, lors qu'on peut encor douter si la notion definie est possible, comme par exemple, si je dis qu'une vis sans fin est une ligne solide dont les parties sont congruentes ou peuvent inceder l'une sur l'autre; celuy qui ne connoist pas d'ailleurs ce que c'est qu'une vis sans fin, pourra douter si une telle ligne est possible, quoyque en effect ce soit une proprieté reciproque de la vis sans fin, car les autres lignes dont les parties sont congruentes (qui ne sont que la circomference du cercle et la ligne droite), sont planes, c'est à dire se peuvent décrire *in plano*. Cela fait voir que toute proprieté reciproque peut servir à une definition nominale; mais lors que la proprieté donne à connoistre la possibilité de la chose elle fait la definition reelle" (A VI, 4B, n. 306, p. 1568). On the topic, see also the letters to Tschirnhaus from May 1678 (A II, 1, n. 167, p. 624) and December 1679 (n. 220, pp. 783–84, again on Hobbes), as well as the one to Malebranche from the 2nd of July, 1679 (n. 207, pp. 721–22).

and construction, is the complete analysis of the notions involved and their resolution into simple ideas. In this case, we directly prove the possibility of the object, without the need of ascertaining its existence; and the definition is called "essential".

> As long as we have only a nominal definition, we cannot be sure of the consequences drawn from it, for if it concealed some contradiction or impossibility, we could draw conflicting conclusions. This is why truths do not depend on names and are not arbitrary, as some modern philosophers have thought. Nevertheless, there is still a great difference between the kinds of real definitions, for when possibility is proved only through experience, the definition is only real and nothing more; as in the definition of quicksilver, the possibility of which we recognize because we know that such a body, extremely heavy and yet rather volatile, is actually found. But when the proof of possibility is presented *a priori*, the definition is both *real* and *causal*, as when it contains the possible production of the thing. And when the definition pushes its analysis back to the primitive concepts without assuming anything which needs an *a priori* proof of its possibility, it is *perfect* or *essential*.[24]

An essential definition may be too difficult to obtain, and it is almost a regulative ideal for mathematical (or philosophical) research. Moreover, it is not strictly necessary to ascertain the possibility of the defined object, as a constructive definition would be sufficient to this effect; if I am able to construct the relevant geometrical object, I cannot doubt its possibility or the consistency of its definition. It should be clear, then, that real and constructive definitions are in fact the main epistemological focus of Leibniz' research.

Strangely enough, one of the sources of Leibniz' preference for constructive definitions may have been Hobbes himself, who suggested employing them in mathematics, even though he was scarcely consistent in this practice, and his alleged proof of the Parallel Postulate is ineffectual just because he fashioned a definition of parallels that itself requires an existential proof.[25] It is hard, in any case, to overestimate the importance of the theory of real definition in Leibniz' philosophy of mathematics, as it merges with a general discussion on constructivism in geometry.

---

[24] The quote is a follow up of the text from § 24 of the *Discours de métaphysique* quoted in the *note* above (A VI, 4B, n. 306, p. 1569; transl. from LOEMKER 1956); in the ensuing correspondence with Arnauld, we find nothing more (cf. Leibniz' letter from July 14th, 1686, A II, 2, n. 13, pp. 65–66). The theory is however repeatedly stated in Leibniz' writings (see for instance another clear presentation of it in the *Meditationes de cognitione, veritate et ideis*, in A VI, 4A, n. 141, pp. 589–90). The distinction between nominal and real definitions may be traced back (with many diversions) to Aristotle himself. Its constructivist application in mathematics is also ancient, and comes again from Proclus' commentary to Euclid (as well as from the commentators of Aristotle in the Late Antiquity); see for instance HARARI 2008. The connection of the real definition with the notion of *essence* is also classical, and was supported by the entire Scholastic tradition (the definition *quid rei*); see Leibniz' remarks on the essence in the previous *note*. We should possibly distinguish between a causal definition and a constructive definition, the latter being just a species of the former (applicable to mathematical concepts only). Also, it is not clear that in the general case an essential definition should also be constructive (even though it may have to be causal), and as an example of a real essential definition Leibniz was probably thinking of the definition of God as *ens perfectissimum* that he thought (against Descartes) to be in need of a (non-contructive) proof of consistency. Note, also, that every geometrical object may have a plurality of constructive definitions, but only one essential definition.

[25] See above, *note* 31 in § 2.

We do not need, now, to consider the whole issue of constructivism in Leibniz' epistemology. It should be clear, in any case, that the internal consistency of a definition is what really matters (for a real definition) and the possibility of actually performing the construction of the desired geometrical figure (at least in this connection) is just a *tool* to prove the consistency of and not an additional constraint (besides non-contradictoriness) on the existence of a mathematical object. Leibniz is not proposing a constructivist epistemology of mathematics in our sense.

Real definitions (in the mathematical case) are, in the end, nominal definitions along with a proof of the existence of their defined objects. The particular status of *mathematical* real definitions over real definitions of empirical objects resides in the fact that in the former case the possibility (and existence) of the defined objects may be proved *a priori*.[26] A very clear statement about the application of Leibniz' theory of real definition to mathematics deals with the correct definition of a circle. Even though several geometrical properties may characterize this figure, a real definition should be given through a genetic procedure:

> All derivative concepts, moreover, arise from a combination of primitive ones, and the more composite concepts from the combination of less composite ones. But one must take care that the combinations do not become useless through the joining-together of incompatible concepts. One must be especially careful, in setting up real definitions, to establish their possibility, that is, to show that the concepts from which they are formed are compatible with each other. So while every reciprocal property of a thing can serve as its nominal definition, since all the other attributes of the thing can always be demonstrated from it, not every such property suffices for a real definition. For, as I have pointed out, there are certain properties which I call paradoxical, whose possibility can be doubted. For example, it can be doubted whether there is a curve for which it is true that given any segment and any point on the curve, the lines connecting this point with the ends of the segment will always form the same angle. … Yet we know that this is the nature of a circle. So, although someone might give a name to the curve having this property, it would not yet be certain that such a curve is possible, and hence that its definition is real. But the concept of the circle set up by Euclid, that of a figure described by the motion of a straight line in a plane about a fixed end, affords a real definition, for such a figure is evidently possible.[27]

---

[26] We note in passing that Kant was to endorse similar views, and this is also relevant for his theory of parallels (for a few references on Kant's theory of parallel lines, see below *note* 10 in § 5).

[27] This is a well-known passage from *De synthesi et analysi universali*, dated 1683–1685 (in A VI, 4A, n. 129, p. 541; transl. from LOEMKER 1956). The same example is given in a letter to Foucher from August 1686 (A II, 2, n. 16, p. 92). The characterization of the circle through the angles in a segment comes from *Elements* III, 21 (although in Euclid it is stated as a property, rather than a characterization proper). Euclid's definition of a circle is that of a plane curve equidistant from a point, and has no constructive marks; Leibniz' reference to him is thus mistaken. Euclid employs, however, a constructive definition of a sphere in *Elements* XI, def. 14, and elsewhere Leibniz acknowledges that Euclid's definition of a circle is "non satis perfecta, quoniam dubitari adhuc potest, utrum talis figura sit possibilis", adding that Euclid's definition of a sphere is better fashioned (LH XXXV, I, 5, Bl. 19r; still unpublished). The kinematic definition of a circle mentioned by Leibniz first appears in HERON, *Definitiones* 27 (even though it is probably more ancient) and was repeatedly discussed since then. Thābit ibn Qurra, introducing for the first time the argument on the generation of a parallel equidistant straight line through the motion of a segment along a line, refers to this definition of the circle (see JAOUICHE 1986, p. 152). In the Renaissance, the constructive definition was used to characterize the circle (among others) by Peletier, who discusses it at length in his 1557 edition of Euclid (p. 6), and was mentioned by Clavius (*Euclidis*, p. 18). It substituted Euclid's definition in

The example of the circle also explains why Leibniz may have considered Clavius' definition of parallel lines through the motion of a segment as a clear example of a real definition to be favored over the Euclidean one given through the obscure mark of non-incidence. Moreover, the definition of parallel lines as equidistant lines, equipped with a complete proof of their possibility is more perfect than the definition through non-incidence, just because from the former one can prove the latter, but not the other way round: from equidistance we immediately deduce non-incidence, but from non-incidence we cannot deduce equidistance without having recourse to external principles (the Parallel Postulate). We see, thus, how Leibniz thought that axioms are provisional principles that are only needed if we don't have good definitions. The epistemological preference for the definition of parallels as equidistant lines, that comes from its constructive character, is explicitly stated by Leibniz in the *Nouveaux Essais*:

> ... the difference is that the real definition displays the possibility of the *definiendum* and the nominal does not. For instance, the definition of two *parallel straight lines* as "lines in the same plane which do not meet even if extended to infinity" is only nominal, for one could at first question whether that is possible. But once we understand that we can draw a straight line in a plane, parallel to a given straight line, by ensuring that the point of the stylus drawing the parallel line remains at the same distance from the given line, we can see at once that the thing is possible, and why the lines have the property of never meeting, which is their nominal definition.[28]

---

some Early Modern editions of the *Elements*, as for instance in Arnauld's *Nouveaux elemens* (ed. Descotes, p. 361) and Borelli's *Euclides* (ed. 1658, p. 19). Borelli, in fact, explicitly complained that it is not evident that a line equidistant to a single point may exist ("non enim facile est videre, an in natura reperiri possit talis figura"; cf. similar statements in *note* 30 in § 2), and thus he made recourse to the constructive definition. In his own copy of the *Euclides restitutus*, Leibniz commented: "rectè". Borelli however also added an axiom stating that all the radii of a circle are equal, since the constructive definition may not warrant this property (this is a fine intuition, and Borelli's axiom may be read, today, as a requirement of isotropy). This definition of a circle is also given as a bright example of a constructive definition by Hobbes (*De corpore*, I, I, § 5; in *Opera*, vol. 1, pp. 5–6). In *Nouveaux Essais*, III, x, § 19, Leibniz gives a similar but more complex example of real definition, discussing two characterizations of a parabola. Here, he seems to ascertain different degrees in the reality (or perfection) of a definition according to different degrees of constructivism: "On peut definir une Parabole au sens des Geometres, que c'est une figure dans la quelle tous les rayons paralleles à une certaine droite sont reunis par la reflexion dans un certain point ou foyer. Mais c'est plutot l'exterieur, et l'effet qui est exprimé par cette idée ou definition, que l'essence interne de cette figure, ou ce qui en puisse faire d'abord connoitre l'origine. On peut même douter au commencement si une telle figure, qu'on souhaite et qui doit faire cet effet, est quelque chose de possible; et c'est ce qui chez moi fait connoitre, si une definition est seulement nominale, et prise des proprietés, ou si elle est encore réelle. Cependant celui, qui nomme la Parabole et ne la connoit que par la definition que je viens de dire, ne laisse pas lorsqu'il en parle, d'entendre une figure, qui a une certaine construction ou constitution, qu'il ne sait pas, mais qu'il souhaite d'apprendre pour la pouvoir tracer. Un autre qui l'aura plus approfondie y ajoutera quelque autre proprieté, et il y decouvrira par example que dans la figure qu'on demande la portion de l'axe interceptée entre l'ordonnée et la perpendiculaire tirées au meme point de la courbe, est toujours constante, et qu'elle est egale à la distance du sommet et du foyer. Ainsi il aura une idée plus parfaite que le premier, et arrivera plus aisement à tracer la figure, quoiqu'il n'y soit pas encore. Et cependant on conviendra que c'est la meme figure, mais dont la constitution est encore cachée" (A VI, 6, n. 2, p. 346).

[28] *Nouveaux Essais*, III, III, § 18; A VI, 6, n. 2, pp. 295 (transl. from REMNANT&BENNETT 1996). See *text 33*.

In Leibniz' foundational qualms about Euclid's definition of parallels we may recognize Borelli's similar concern that a definition having recourse to the *ignota aut dubia* property of infinite extension cannot show the possibility of its object. From Leibniz' many attempts for producing the Parallel Postulate, however, we also know that he was not following Borelli halfway, and that both men realized that even the clear constructive procedure for producing an equidistant straight line has to be grounded in some deeper principle. It is not obvious that the line drawn following Leibniz' precept, which is equidistant to the given straight line, will be itself straight. Leibniz' stance in the *Nouveaux Essais* could not be his last word on the subject: real definitions, however evident, themselves require a proof.[29]

We enter here the more secluded world of a full-fledged *analysis situs*. A constructive definition may guarantee the possibility of its object (an equidistant line, for instance), but not yet its properties (to be straight). It may satisfy the imagination, which is able to produce *a priori* the required ideal object, but not the understanding that needs a firmer ground to build an unhypothetical science. Leibniz' call for constructive definitions is to be taken as a good epistemological request to secure the consistency of most mathematical notions, but the ultimate foundation of geometry needs something more: it requires grounding and explaining the very act of construction by a thorough analysis of the conditions of possibility of geometrical objects. At the culmination of his work on *analysis situs*, Leibniz has to try to proceed beyond the theory of constructive definitions to reach the level of an *essential* situational analysis of the geometrical concepts.

---

[29] If we accept the idea that a real definition should be proved, one might object to Leibniz that Euclid's definition of parallels as non-incident lines is also proved as real through the construction of parallels in *Elements* I, 31. Thus, there is no point in changing the original Euclidean definition in the hope of providing another one whose possibility should be more evident. This objection has been actually raised, in fact, by Saccheri against Borelli's new definition of parallel lines (see *Euclides vindicatus*, Scholium 2 after Proposition 21, pp. 31–32), and again by Lambert against Wolff (*Theorie der Parallellinien*, § 3, pp. 139–40; but Lambert might have in mind Leibniz himself, as the *Nouveaux Essais* were published in 1765 and Lambert's theory of parallels was written the following year). Leibniz' answer to these concerns might be a simple issue of methodology, stating that you have *first* to prove the reality of the definitions and *then* you may proceed to the *theorems* in the proper sense (so that the proof of the possibility of the defined object is not, properly speaking, a theorem); this would be an answer after the Aristotelian fashion, based on the idea that there is no science if you are not starting from well-defined objects (an inconsistent set of definitions); for similar remarks in Kant, see the above-mentioned paper by Heis [A]. However, Leibniz might also have answered (more to the point) that certain definitions are not capable of being transformed into *essential* definitions, and Euclid's original characterization of parallel lines as non-incident straights cannot capture the *essence* of parallelism. Such reference to essence (which may sound an over-metaphysical notion, even more since Leibniz has stated his own mild nominalism and claimed that every definition is in principle equally good, provided that it is consistent) should be framed in Leibniz' general theory of a geometry of space, and on the possibility of grounding the notion of equidistance (but not the notion of non-incidence) on the basic *situational relations* and the definition of space. In other words, an essential definition should show the *simple* ideas constituting the notion of the defined object, and the concept of *distance* (in the definition of parallels as equidistant lines) is one of such ideas, while the concept of non-incidence is (for Leibniz) derivative and more complex. We are going to see some of these developments in the following pages.

In other words, the above-mentioned definition of parallel lines as equidistant lines is *real* and *constructive*, but it is still not *essential* (or *perfect* in a strong sense) as we have to *assume* some principle (the Parallel Postulate, in fact) to warrant the possibility of the construction itself. The same may be said about the constructive definition of a circle with respect to the *Elements*: Euclid does not prove that a circle is possible, but postulates the possibility of its construction (through the Third Postulate). The pseudo-Euclidean definition of a circle through the rotational flow of a point is thus real and constructive, but not yet essential; to have an essential definition, one should prove the Third Postulate itself.[30]

The quest for a system of essential definitions is thus Leibniz' way to the foundations of geometry, and the major aim of an *analysis situs*. It requires the proofs of all the Euclidean principles (including the Parallel Postulate) through the investigation of the condition of possibility of all geometrical objects and all their possible constructions. The ground of the possibility of all geometrical objects and constructions, finally, is to be found in *space itself*, its properties, and its structure. The analysis of situations becomes in the end a science of space.

3.3 Geometry as the Science of Space.

It is quite straightforward that in the context of seventeenth-century mathematics, in which axioms are derivative from definitions, the request for a proof of the possibility (or existence) of the defined object is analogous to the modern request to prove the consistency (or the existence of a model) of a system of axioms (that now we take as more fundamental than the definitions). There are obviously a few important differences that should be taken into consideration. These, in turn, may lead to some simplifications or misunderstandings that naturally arise when taking at face value Leibniz' theory of real definition.

First of all, it seems that the theory of real definition could somehow provide an *implicit* axiomatic basis for geometry. One may take the definition of parallel lines as equidistant straight lines (an assumption that we know to be equivalent to the Parallel Postulate), and show that it is a real definition. This shouldn't be too difficult, of course, as we know today that classical Euclidean geometry is consistent, and thus equidistant straight lines are definitely possible. In this way, one arrives in fact at an axiomatization of Euclidean geometry through a set of real definitions. Following this path, however, Leibniz' theory seems to actually open toward a plu-

[30] In general, postulates (as distinguished from axioms) are just *assumptions* of the possibility of geometrical objects: "Postulata: hoc nihil aliud sunt quam assumtae possibilitates, aut generationes" (from some *Definitiones* in LH XXXV, I, 5, Bl. 19v, quoted by ECHEVERRÍA 1992, p. 38; cf. our *text 12*). On the circle in particular, see: "Les Geometres, qui sont les veritables maistres dans l'art de raisonner ont vû que pour que les demonstrations qu'on tire des definitions soyent bonnes, il faut prouver, ou postuler au moins, que la notion comprise dans la definition est possible. C'est pourquoy Euclide a mis parmi ses *postulata*, que le cercle est quelque chose de possible, en demandant qu'on puisse decrire un cercle dont le centre et le rayon soyent donnés" (a draft for a letter from 1700; in GP IV, p. 401).

rality of geometries. Since both the definition of a Euclidean parallel line (given, for instance, through equidistance) and a hyperbolic parallel line (given, for instance, through asymptotic properties) are (each) internally consistent, it would seem that Leibniz had to admit the possibility of building a system of non-Euclidean geometry. The theory of real definition, as it is only concerned with the internal consistency of the notions involved, would not imply that only *one* geometrical system is possible. Additional logical or epistemological principles are needed to pass from the theory of real definition to the Leibnizian claim that there is only one possible geometry.

A further difficulty with this approach is that Leibniz' requirement for the internal consistency of each definition falls short of a demand for the consistency for the whole geometrical system, where the definitions (or rather the axioms, from our modern point of view) should be not only internally consistent one by one, but also as a whole. Leibniz' point of view seems (at first) to be in line with the general development of classical geometry, which is a geometry of *figures*. These figures are defined one by one, and their reality is also conceived as that of independent beings. In classical geometry there is no such thing as a *space* in which figures are drawn and whose features determine the figures' properties. Space, in fact, would be characterized and captured through a set of axioms that establish relations among basic elements and, in so doing, define a structure. On the other hand, a collection of individual and unrelated figures (a straight line, a circle, a cone, a sphere) are aptly conceptualized through a set of definitions, and this is the case with geometry from Antiquity to the Renaissance. To use Cassirer's words, we might speak of a geometry of substances as opposed to a geometry of functions (or structures).[31] The obvious drawback of the ancient perspective (at least for our present matter) is that as soon as geometry develops enough to open toward the possibility of different geometrical systems, as is the case with the seminal investigations on the Parallel Postulate, the internal consistency of a set of definitions *separately* (that is, *distributive*) may not be enough to guarantee the overall consistency of the system (as a *collective* consistent set). A system of axioms (disguised as definitions) may well be inconsistent even though each axiom is not contradictory in itself. In other words, Leibniz' theory of real definitions seems incapable of overcoming the difficulties that he found in Hobbes' alleged super-nominalism; it is still open to contradiction.

Leibniz' epistemology of geometry is however developed and complex enough to accommodate the above-mentioned quandaries. As a matter of fact, Leibniz' *analysis situs* goes far beyond classical geometry or the epistemology of real definition. The main accomplishment of Leibniz' geometrical studies (I would say) is precisely to begin to turn geometry from a theory of figures into a *science of space*. Already in his first attempts at an *analysis situs*, Leibniz was crystal-clear that the objects of

---

[31] The obvious reference is to Cassirer's *Substanzbegriff und Funktionbegriff* (now in CASSIRER, *Werke*, vol. 6), whose Chapter 3 is dedicated to these concepts in mathematics. Cassirer insisted on the role of genetic procedures in modern geometry (that he called "serial concepts") in opposition to the static figures (allegedly) employed in ancient mathematics. This distinction had probably a lesser impact on the history of epistemology than the one that Cassirer attributed to it, but nonetheless he was able to clearly envisage that the crowning principle of modern geometry was the conception of space as a structure (cf. esp. §§ 3–4 of the Chapter).

geometry are not figures or continuous magnitudes (as in the Greek tradition), but space itself. This is an important conceptual revolution indeed, even though Leibniz' geometry is not always equal to his epistemology.

In the seventeenth century, the dominant views about the ontology of geometrical objects took *material* extension as their substrate; Descartes may be naturally mentioned in this connection, but similar theories were held in the Scholastic tradition as well. A few geometers, however, were already conceiving of space as the subject-matter of geometrical investigation. For instance, Francesco Patrizi's "new geometry" of spatial figures, or Newton's idea that geometrical objects are sections and parts of an all-encompassing absolute space, already testified to such a change in ontology.[32] Nonetheless, these metaphysical conceptions of space still regarded it as a geometrically amorphous background which is a mere ontological precondition of the possibility of geometrical figures (which are parts and pieces of it). Neither Patrizi nor Newton (nor anyone else) regarded space as something endowed with geometrical properties. Geometry was still a science of *figures in space*, not a science investigating the properties of space itself. The important step forward taken by Leibniz in the ontology of geometry was to consider it as a structure.

Leibniz' celebrated definition of space (the one he was to employ to fight against Newton's conception of an absolute space) is that of an *order of situations*.[33] This metaphysical definition is often discussed in the context of an appraisal of Leibniz' ideas about ontology or physics; it is (correctly) regarded as the foundation of Leibniz' mechanics, and the dispute between Leibniz' and Newton's notions of space

[32] Patrizi's new metaphysic of space was first expounded in his two celebrated booklets *De spacio physico* and *De spacio mathematico* from 1587 (later collected and reprinted in Patrizi's main metaphysical work, the *Nova de universis philosophia* from 1591); but he also wrote a book on a *New Geometry* (*Della nuova geometria*, 1587), that is a geometry whose object is space itself: "Le Matematiche tutte, e principali, e subalterne, ne si astraggono dalle cose naturali, ne sono nella fantasia, ne nella dianea, ma lo spazio è generale lor subietto" (*Nuova geometria*, p. 2). Patrizi's ideas enjoyed a wide circulation in the seventeenth century, and Newton's own metaphysical conception of space is probably dependent on them (through Gassendi or the Cambridge Neoplatonists, who in turn read and commented on Patrizi's original work). As for Newton's conception of the objects of geometry, see for instance: "Et hinc ubique sunt omnia figurarum genera, ubique sphaerae, ubique cubi, ubique triangula, ubique lineae rectae, ubique circulares, Ellipticae, Parabolicae, ceteraeque omnes, idque omnium formarum et magnitudinum, etiamsi non ad visum delineatae. Nam materiali delineatio figurae alicujus non est istius figurae quoad spatium nova productio, sed tantum corporea representatione ejus ut jam sensibus apparet esse quae prius fuit insensibilis in spatio" (*De gravitatione*, p. 100; ed. Hall). Leibniz had read Patrizi's metaphysical writings, and even his *Nuova geometria*, which he regarded (correctly) very weak as a mathematical book but very important in epistemology (see *Projet et essais pour avancer l'art d'inventer*, in A VI, 4A, n. 205, p. 966). Patrizi, among other things, had attempted a purely logical exposition of the most elementary Euclidean theorems – even though his actual results in this direction fell shorter than those by Herlinus or Clavius. On Patrizi's new geometry, see De Risi 2014 [B].

[33] A celebrated quote from the Leibniz-Clarke exchange: "Je ne dis point donc que l'espace est un ordre ou situation, mais un *ordre des situations*, ou selon lequel les situations sont rangées; et que l'espace abstrait est cet ordre des situations conçues comme possibles. Ainsi c'est quelque chose d'ideal …" (*Fifth paper to Clarke*, § 104; in Robinet 1957, p. 171). The definition appears often, especially in the writings on *analysis situs*. Among the recent contributions to Leibniz' theory of space, see Futch 2008 and Arthur 2013B.

turns out to be a general discussion on the meaning of inertia, forces, and relative motion. It should not be overlooked, however, that Leibniz' definition of space as an order of situations is first of all a purely *geometrical* claim about the mathematical structure of space. It is in fact the first conception ever of a geometrical space as a structure (while Newton's *geometrical* space is still unstructured). The very concept of situation, which is the cornerstone of Leibniz' definition of space, is a geometrical notion: the one investigated in the *analysis situs*.

In geometrical terms, a *situation* is a relation between objects. Even though we may want to consider situational relations between extended geometrical figures (such as straight lines or circles), the ultimate objects of the relation of *situs* are called *points*. Note that in this very abstract sense, a point is nothing more than an object which *is conceived* as only a *relatum* of a situational relation.[34] Space itself is a set of points and of the relations among them. This peculiar relation of *situs* is left undefined, but it is characterized through distance: two sets of points have the same reciprocal situations if and only if they are congruent (isometric).[35] For instance, the three vertices of two triangles with equal sides, or the points of two circumferences of equal radius, or the ends of two segments of the same length, have the same reciprocal situation. In this way, the definition of space as an order (or system) of

---

[34] The very abstract conception of a point as a relational element comes to Leibniz from several sources, but the elaboration and centrality of the notion is his own. Hobbes stated that a point is not unextended but what is *considered* as such: "… per punctum non intelligi id, quod quantitatem nullam habet, sive quod nulla ratione potest dividi (nihil enim est ejusmodi in rerum natura); sed id cujus quantitas non consideratur, hoc est, cujus neque quantitas neque pars ulla inter demonstrandum computatur; ita ut punctum non habeatur pro indivisibili, sed pro indiviso" (*De corpore*, III, XV, § 2, in *Opera*, vol. 1, p. 177; cf. *Six lessons*, in *The works*, vol. 7, pp. 216–17). Other authors insisted on the concept of *situs* to characterize a point. The prehistory of this conception is to be found again in Aristotle who defines the point as a monad with position (μονὰς θέσιν ἔχουσα; see *De an.* A 4, 409a5-7; *Metaph.* Δ 6, 1016b24-29; *An. post.* A 27, 87a35-37), but the modern geometrical developments were in Francesco Patrizi's *Della nuova geometria* (pp. 14–15), in Pascal's *Introduction à la géométrie* (*Oeuvres*, vol. 9, p. 292), or in Mercator's edition of the *Elements* (p. 3), all of them well known to Leibniz. Here is one of Leibniz' formulations: "'Punctum est cujus pars nulla est'. Addendum est, situm habens. Alioqui et temporis instans, et Anima punctum foret" (*In Euclidis* πρῶτα, in GM V, p. 183; the quote refers once more to Clavius' commentary, where the Jesuit mathematician had written: "… ut nimirum excludemus animam rationale, nunc vel instans temporis, & unitatem, quae etiam partes non habent"; the reference to *situs* is however missing in Clavius). As for the relation with space: "Spatium est ordo coexistendi … Situs est coexistendi modus" (*Initia rerum mathematicarum metaphysica*, GM VII, p. 18).

[35] Leibniz used a definition by abstraction, that he also famously employed to define the (situational) concept of place (*locus*) in the correspondence with Clarke (*Fifth Paper* to Clarke, § 47; in ROBINET 1957, pp. 142–45). See similar passages in the *De realitate accidentium*, in A VI, 4A, n. 209, p. 996; and *Demonstratio omnimoda*, in DE RISI 2007, p. 602. The source of this kind of definitional procedure was Euclid's (or rather Eudoxus') alleged definition of "ratio" through the definition of equal ratios (proportion) in *Elements* V, def. 5. That in defining proportion Eudoxus wanted to define ratio is obviously highly debatable (*Elements* V, def. 3 is itself a straightforward, even if slightly "philosophical", definition of ratio). But some Early Modern systematizations of the classical theory of proportions (like the one by Commandino) suggested in fact that *Elements* V, def. 3 was just a characterization of the real, operational definition given in *Elements* V, def. 5. On the topic, see GIUSTI 1993.

situations is roughly equivalent to the modern concept of a metric space, which is given by a set of points and a set of distances between them. A space is just a system of distances – a *metric structure*.

Leibniz' *analysis situs* is thus mainly a theory of metric spaces (and their subspaces). He was able to envisage and define several very abstract properties that these spaces may or may not possess. Among them, we may mention *uniformity*, which is a quite good approximation to our notion of isotropy; or *homogeneity*, which is linked to our modern conception of a topological manifold; or *continuity*, of which Leibniz is able to give a very good characterization (approaching Dedekind's completeness).[36] The *analysis situs* is thus designed to define and mathematically treat spaces with these (and other) features. Leibniz, for the first time in history, was thus able to assert (provisionally, in his view) axioms about space itself, and to prove theorems not about lines or circles, but about abstract space-structures. If we situate the divide between Ancient and Modern geometry in the transformations of this science from a theory of figures into a theory of structures, then we should credit to Leibniz' *analysis situs* the dignity of the first experiment ever in this direction.

Leibniz' point was however much stronger than this. In his writings on the *analysis situs*, in fact, he introduced the notion of an *absolute space*. This has nothing to do with Newton's homonymous concept, nor with mechanics or inertial frames. Leibniz' *spatium absolutum* (or *integrum*, or *universum*) is a total order of all possible situational relations. It is an all-encompassing geometrical space, and every other situational order should be considered as a proper subspace of it (a subset of its points and relations).[37] As such, this absolute space is the actual ground of the possibility of all the other geometrical figures. Circles or parallel lines depend for their possibility on the possibility of being embedded in, and being part of, absolute space. Thus, the entire geometrical construction of Leibniz, in the end, hinges on the properties of absolute space, and his nominalism extends just to the point of accept-

[36] For the definition of these concepts in Leibniz' geometrical writings, see De Risi 2007, pp. 178–214. Leibniz' definitions of continuity as linear completeness are especially good in the *Specimen geometriae luciferae* (GM vii, pp. 284) and the *In Euclidis* πρῶτα (GM v, p. 184).

[37] Some references on the *spatium universum*, *integrum* or *absolutum*: "Def. Spatium Universum est locus omnium punctorum" (*Scheda*, 1695; in De Risi 2007, p. 588). Moreover: "Spatium integrum est locus omnium punctorum. Itaque si quodlibet punctum vocetur X erit spatium $\bar{X}$", where the symbol $\bar{X}$ means the collection of all X's (*Magnitudinis nomine*, 1712; in De Risi 2007, p. 596); cf. also § 89 of *Characteristica geometrica* (in Echeverría 1995, p. 220). Finally: "Spatium absolutum interminatum est quod in situ amplissimum est. Ideo omnis locus est in eodem spatio absoluto. Spatium absolutum est continuum, alioqui locus aliquis interponi posset qui in ipso non esset. Spatium absolutum uniforme est, varietas enim orietur à determinationibus, quae cum amplissimo non consistunt." (*Spatium absolutum*, 1714; in De Risi 2007, p. 609). Leibniz would like to *prove* that there is an absolute space in this sense (he tries this, in fact, in the above-mentioned *Scheda* from 1695), starting from the idea that everything that has a situation, has a situation with respect to everything else: "Ea est natura situs, ut omnia quae habent situm, habeant etiam situm inter se" (*Prima geometriae principia*, in C 541). Thus, it is impossible to have different situational systems (spaces) that are not situated among themselves, and are not parts of a bigger encompassing space. We should remark that among the more recent attempts at a foundation of geometry, Veronese had a similar concept, that he called "*spazio generale*" (Veronese, *Fondamenti di geometria*, pp. 210–11).

ing arbitrary definitions of sets of points that can be embedded in a three-dimensional real Euclidean manifold.

That Leibniz may not have been perfectly aware of the novelty of his approach is testified to by his reliance on the above-mentioned theory of real definitions as the true principles of geometry: a theory which is more apt to ground classical geometry than a science of space. Nevertheless, the theory of real definition acquires other meanings in this new perspective. In fact, since every figure is seen as a particular system of situational relations, if we analyze deeper and deeper every notion involved in a definition, as Leibniz requires us to do in order to arrive at an essential definition which expounds all the simple notions that constitute the concept of the object, at some point we will arrive at the concept of *situs* itself and we will represent the geometrical object as a configuration of distances. These are thus the proper "essential definitions" of Leibniz' *analysis situs*, and they are formally expressed in the symbolism of a *characteristica geometrica*.[38]

Since a similar analysis may be carried out for every definition of the figures employed in geometry, the internal consistency of each geometrical notion will be linked (through the concepts of situation and space) to the internal consistency of every other geometrical notion. In fact, every possible figure has to be embeddable in the same absolute space, which is now the condition of the *compossibility* of the various geometrical objects.[39] The introduction of the notion of absolute space acts as a further constraint on the geometrical theory, whose objects have to be both internally possible and externally compossible with other embedded objects in space. It may be important to stress that this compossibility of geometrical objects cannot be evaluated through *constructive* real definitions, which are only concerned with the *internal possibility* of the defined objects; these definitions are so important in classical geometrical theory (and for Leibniz as well), because they are in fact adequate to establish this kind of possibility. But to ascertain the *external compossibility* of figures in space one has to resolve their definitions into the basic situational notions that characterize them as metric subspaces and systems of distances; In short, Leibniz' deeper conception of geometry as a science of space requires us to proceed toward

[38] Think, for instance, of a circle as a system of equal distances from the center: this is Leibniz' formal definition of it in his writings on *characteristica geometrica*, and it only employs situational relations. If A is a point in a plane, and A.X is the situation (distance) of the point X with A, then the locus of all X's that are at the same distance from A as another given point B, is a circle (of radius AB). This is expressed by $A.B \simeq A.X$, a formula that only relies on the symbols to express points (abstract situational elements), distances, and the fundamental relation of congruence ($\simeq$). Some further examples will be given in § 4.4.

[39] Leibniz' standard notion of compossibility only concerns the possibility of things existing in the same world (e.g. two monads are singularly possible if each of their complete notion is not contradictory, and they are compossible if they can exist in the same possible world). Space, on the contrary, is more often qualified as an order of *coexistence*: see for instance *Initia rerum mathematicarum metaphysica*, in GM VII, p. 18. In sum, Leibniz usually does not employ his famous notion of *compossibilitas* to characterize the existence of different mathematical entities in the same space. There are however a few passages in which he actually talks about compossibility of defining space itself (see for instance his letter to De Volder from June 20th, 1703; in GP II, p. 253). In any case, the concept (if not the word) is clearly represented in his geometrical writings.

essential definitions. In this way, however, the set of definitions acts as a *system* in the proper sense, whose consistency can be evaluated only collectively and not from the definitions one by one. We have thus recovered a more modern meaning of the consistency of a geometrical system.[40]

A reference to the concept of an absolute space is also useful to *ground* constructive definitions. For instance, the nominal definition of a circle as a curve defined by the above-mentioned angular property (whose possibility we might doubt) is aptly replaced by the real definition stating that a circle is the curve traced by the free end of a segment rotating around a fixed point. In several papers, however, Leibniz saw the necessity to found the possibility of such a rotational movement as well, and to ground it in the structure of space itself (in particular, in its uniformity, i. e. isotropy). In discussing the principles put forward by Arnauld in the *Nouveaux elémens de géometrie*, Leibniz commented that we need a proof of Euclid's Third Postulate grounded in the properties of space:

> *First postulate* on the circle, that we can describe it in any given point and with any given radius: to do this, we have to assume that a straight line can be moved in a plane while an extremity of it remains fixed. *Remark.* This seems to me a consequence of the uniformity of space with respect to a sphere, and the uniformity of the plane with respect to a circle. A point in space or plane, in fact, has the same relation with all the sides. … *Third axiom.* A circular line is uniform throughout and thus the circumferences with the same radius are equal. *Remark.* This does not follow from the uniformity of a circle, but from the nature of position. …[41]

We see here the necessity to push to the end the analysis of the basic notions involved in geometry, to get not only a real definition of the circle, but also an *essential* definition of it. Since space is the ultimate logical and ontological source for the figures investigated by classical synthetic geometry, the complete analysis of all geometrical

---

[40] The classic book by Russell on the philosophy of Leibniz (from 1900) famously claims that many of the shortcomings in Leibniz' philosophy depend on his reliance on the Aristotelian ontology (and logic) of subject and predicate (which may be paradigmatically expressed in a theory of definitions as the foundation of science). Already Cassirer's essay on *Leibniz' System* from 1902 (now in Cassirer, *Werke*, vol. 1) judged that Leibniz was in fact trying to go beyond the classical ontology and was projected toward a theory of functions, structures and relations. This is especially true, in fact, for his writings on *analysis situs*, that Cassirer discusses, and Russell ignores. Nevertheless, it is true that, even in his most daring geometrical constructions, Leibniz remained somehow entangled in the classical views, and he wavers between old and new concepts of geometry. Moreover, he saw the necessity of a logical treatment of relations, but lingered in an improved syllogistics that fell short of it. The most complete and historically accurate treatment of Leibniz' theory of relations is Mugnai 1992.

[41] The above text can be found among Leibniz' personal notes to Arnauld's work. It was partially edited in Bopp 1929, and in our *text 22* I give a transcription of the original Latin passage. Arnauld's postulates are from Section 2 of the Fifth Book of the *Nouveaux elémens*: "*Premier axiome ou demande.* On demande qu'ayant un intervalle donné, on puisse décrire une circonférence de cet intervalle. Ce qu'on ne peut douter être possible, puisqu'il ne faut pour cela que concevoir que la ligne qui joindra les deux points de cet intervalle se remue, l'une de ses extrémités demeurant immobile. … *Troisième axiome.* La manière dont l'on conçoit que se forme la ligne circulaire est si simple, qu'il est impossible de concevoir qu'elle ne soit pas partout dans une entière uniformité. Et de là il s'ensuit que les thórèmes suivants sont naturellement connus. Les circonférences qui sont décrites d'un égal intervalle sont égales. …" (*Nouveaux elémens*, pp. 363–65; ed. Descotes).

concepts (that is the main aim of an *analysis situs*) which allows the transition from *real and constructive* definitions to *properly essential* ones must be grounded in the structural properties of space (such as uniformity), and in the very notion of position (i. e. situation, *situs*).[42]

The same may well happen with the definition of equidistant straight lines given by Clavius through the motion of a segment along a line: this definition is constructive, and thus might prove the possibility of equidistant lines; but Leibniz acknowledged that it is not truly sufficient for an essential real definition, as one has to demonstrate the possibility of that kind of movement first, and this may only be done through an analysis of the properties of the space itself in which the parallel line is described. Thus, Clavius' proof of the Parallel Postulate which was rested on this definition, is far from being sufficient, and Leibniz' appeal to constructive procedures has to be integrated with a theory of absolute space as a condition of possibility for such constructions.

It is important to stress that this concept of an absolute space is not an empirical or given concept; even though Leibniz was later to claim that physical space is just a realization of it, it is not initially conceived of as physical space. In Leibniz' view, it is simply the geometrical concept of the total order of situations. It would be interesting to follow Leibniz in his attempts to ground the possibility of absolute

---

[42] We have several other writings on *analysis situs* in which Leibniz claims that the possibility of a rotation that generates a circle (and a straight line as an axis) has to be proved and cannot be assumed. See for instance: "Generatio quidem rectae et circuli habetur nullo jam recta et circulo suppositis, nempe si corpus aliquod moveatur duobus suis punctis immotis. Punctum quodcunque corporis motu suo describet circulum; omnia autem puncta quiescentia cadent in rectam. Attamen ut res non experimento atque imaginatione, sed sola mente sive demonstratione agatur, demonstrandum est corpus aliquod seu punctum saltem in eo moveri posse duobus punctis corporis immotis; demonstrandum quoque est innumera alia puncta quiescentia in unum continuum cadentia in eodem corpore posse dari" (*Generatio quidem rectae et circuli*, around 1676; in Echeverría 1995, p. 66). Leibniz resorts to the *uniformity* of space to ground this kind of rotational flow, for instance, in his *Uniformis locus* from 1692 (in De Risi 2007, pp. 582–85). See also the *In Euclidis* πρῶτα: "Postulat. III. Quovis centro et intervallo circulum describere. Id in plano efficit motus radii uno puncto immoto. Posse autem moveri rectam uno puncto immoto ex eo colligitur, quod spatium planumve uniforme est, et quod versus unam est plagam, potest etiam versus aliam sumi quamcunque" (GM v, p. 206). We find similar statements, in the same essay, about the other constructive postulates of the *Elements* (in Clavius' version). It should be remarked, finally, that in one single fragment Leibniz reduces the possibility of the constructions of geometrical figures to the construction of straight lines and circles (as in the *Elements*), and the latter to the constitution of space, body and motion (and not to space only), while these three depend on God: "Ita omnes lineae motuum in tota Geometria revocantur ad duos tantùm motûs, unum in linea recta alterum in linea circulari. His duobus enim suppositis demonstrari potest alias omnes lineas, exempli causa, Parabolam, Hyperbolam, Conchoidem, Spiralem, possibiles esse. Rectam autem duci et circulum describi Euclides non docuit, sed postulare satis habuit. Quamquam posito spatio, corpore, linea recta et motu continuo, possit etiam demonstrari possibilitas circuli. Imò et linea recta demonstrari potest posito spatio et corpore et motu continuo. Quid autem de tribus his continuis sentiendum sit videtur pendere ex consideratione perfectionis divinae. Sed Geometria ad haec assurgere necesse non habet" (C, pp. 431–32). We note that Borelli's Axiom 12, that all the radii of a circle are equal, may have been a source for this kind of consideration (see above, *note* 27); in Borelli, however, there is no reference to space.

space, and thus prove that its definition as a total order of situations is in fact a *real* (and essential) definition – the real definition that grounds all other real geometrical definitions. These attempts, however, are much more metaphysical than geometrical and are related to Leibniz' general theory of space as a kind of form of sensible perception (an object that is not properly real or substantial, but "contains something ideal", as Leibniz says), and to the deepest results of his late idealistic metaphysics. We just mention that Leibniz thought he had philosophical arguments to support the claim that a total order of situations is possible, and in fact real (a phenomenological object), as a condition of perception. He even arrived at envisaging an actual *construction* of it as a product of (distinct) imagination.[43] We may also add that this is the reason why Leibniz claimed that geometrical ideas are *innate*. This kind of innatism should not be taken at face value as a naïve psychological thesis about people having ideas of every mathematical object from birth. We have seen, in fact, that geometrical concepts are arbitrarily fashioned and are thus *made* by mathematicians, not *given* to them either *a posteriori* (through abstraction) or *a priori* (through introspection or reflection). The building blocks of these concepts, and the notion of *situs* in particular, are however grounded in the basic phenomenological operations that any mind is able to perform to explore the external world. Space itself is a necessary (and "innate") product of these intentional acts.[44]

[43] These considerations about the production of space by the perceiving subject are to be found in the last sections of the *Characteristica geometrica* from 1679. These last §§ 104–108 were not published by Gerhardt in the classical edition of Leibniz' mathematical works, and have been rediscovered by Echeverría and edited by him (in Echeverría 1995, pp. 226–32). Something similar, however, may be extracted from the *Initia rerum matematicarum metaphysica* from 1715, showing that Leibniz continued to think on the matter all his life (see Gm vii, pp. 17–29). Note, at least, that Leibniz' main axiom that everything which is situated is also situated with respect of everything else, which is the foundation of absolute space (see the previous *note* 37), is consistent with the idea that a situational relation is a feature of perception, so that a plurality of spaces would mean a plurality of uncommunicating spatial experiences of the same monad (a divided perceptual consciousness). In a quite speculative reconstruction of Leibniz' spatial idealism, I have tried to connect these passages with the problem of a determination of the spatial metric and thus to the Parallel Postulate itself (cf. De Risi 2007, pp. 405–28). Given the purely metaphysical character of these reflections, however, we will not discuss them here.

[44] In fact, Leibniz claims that mathematical notions and their analytical consequences are perfect exemplifications of *innate ideas*: Were someone to ask him, he writes, which truths are innate, "je lui nommerois les propositions d'Arithmetique et de Geometrie, qui sont toutes de cette nature, et en matiere des verités necessaires, on n'en sauroit trouver d'autres" (*Nouveaux Essais*, I, i, § 23; A vi, 6, n. 2, p. 86). More in particular, we have already seen that the notion of *situs* is defined by abstraction (see above, *note* 35), making reference to *congruence*. Congruence itself is analyzed by Leibniz in the two basic geometrical concepts of similarity and equality (two things being congruent to each other if and only if they are equal in measure and similar in shape), while equality and similarity receive, in turn, a purely phenomenological characterization: similar things (for instance) are those I cannot distinguish if I perceive them separately (see below, *note* 48 in § 4). Our ability to recognize the identity in shape of different things, which grounds Leibniz' conception of geometrical similarity, is not conceived by him as a simply psychological and accidental feature of our human mind, but as a fundamental phenomenological ability that grounds every monad's perception of the world. The

More than this, Leibniz thought that such an absolute space is completely determined as regards its properties. Absolute space is, in fact, uniform, homogeneous, continuous (complete and connected), and (besides this) also tridimensional and Euclidean. The idea is thus that from the very notion of a complete system of situations, one should be able to derive all these very characterizing properties of space. In this sense the ultimate ground for Leibniz' proofs of the geometrical axioms should be the simple concept of a "total" metric space (whatever "total" may mean in this connection). There is just one absolute space, and thus geometry is one as well. From the metaphysical point of view, Leibniz may express this by claiming that space is the same in all possible worlds, as it cannot be otherwise. It is an ideal entity, and as such it is unaffected by the contingent properties of the existent substances of the world (the monads):

> Mass and its diffusion result from monads, but not space. For space, like time, is a certain order, namely (in the case of space) that of coexisting, which includes not only actual things but also possibles. … If there were in the nature of things other subdivisions of organic bodies into organic bodies, there would be other monads, other mass, and yet the space filled would be the same. [45]

We will not, in the present study, attempt to trace out all of Leibniz's attempt to prove all the features of absolute space. It is obviously a grand and visionary program, yet doomed to complete failure. It mixes mathematical investigations with philosophical reflections, and represents the true and ultimate aim of Leibniz' work on the foundations of geometry (and the metaphysics of space), as well as the crowning result of the whole *analysis situs*. It is important to understand, however, that Leibniz' proofs of the Parallel Postulate were devised precisely in order to advance toward this goal.

3.4 Philosophy and the Parallel Postulate.

It should be remarked that the latter approach to the foundations of mathematics, in which Leibniz attempted to prove the possibility of geometrical objects through their reference to space, has the relevant advantage of understanding the Parallel Postulate itself as expressing a property of space. This was not the case, of course, in classical geometry, where the Postulate was regarded as a statement about straight lines and

faculty to perform this cognitive act is thus *innate* in us in a strong sense; by reflecting on this act we form the concept of *situs*, and from it we arbitrarily fashion the other complex geometrical notions.

[45] Leibniz' letter to Des Bosses from July 9th, 1709; in GP II, p. 379 (trans. from LOOK&RUTHERFORD 2007, p. 141). We cannot discuss the relations between mathematical and physical space in Leibniz' thought, as this would require almost an entire evaluation of his metaphysics. We may just remark that Leibniz' definition of space as an order of *possible* situations (see above the quote in *note* 33) already hints at the idea that every possible world will have the same space-structure. The same direction is taken by the entire Leibnizian epistemology of mathematics, which always takes the geometrical theorems as *necessary* truths (i.e. proposition true in every possible world). For a contrary opinion, see BELAVAL 1978 and RESCHER 1981.

their incidence – in a geometry of figures. The understanding that the theory of parallels is implied in the structure of space itself is indeed an important and non-trivial advance toward the creation of non-Euclidean geometries.[46]

It is true that Leibniz was not as explicit on this point as he is with other space-properties. He had no technical term to express the flat metric of space, as he had for isotropy or completeness; and he was not always adamant that the Parallel Postulate is in fact one of the most important foundational principles for a geometry of space. Sometimes, he was even confused enough to think that isotropy is a sufficient condition for flatness. Other times, however, he was prompt to distinguish the two notions, and realized that a spherical surface, for instance, is isotropic without being Euclidean (i. e. flat). In these contexts, Leibniz often defined Euclidean space and its flat subspaces through the property of self-similarity, which amounts to the possibility of the transformations by similarity of their parts.[47]

This last characterization of the Euclidean space is very important. In the nineteenth century it was to come to be called *isogeny*, and many were to attempt to ground the (supposed) flatness of physical space on this property.[48] Clearly, Wallis' proof of the equivalence between the existence of similar figures and the Parallel Postulate paved the way for considering this Euclidean principle as a structural property of space and not a feature of lines, and encouraged philosophers and mathematicians not to regard the attempts to prove the Postulate as a mere methodological concern in the field of elementary geometry. Wallis' proof, in other words, opened up the possibility of regarding the Parallel Postulate as an important metaphysical principle that governs the behavior of space and spatial objects. Leibniz was probably the first to discuss the matter from both a mathematical and a metaphysical point of view. He put forward a famous thought experiment about a proportional reduction or growth of the whole cosmic space, and claimed that such a change in size of the entire universe would go completely unnoticed, since we cannot perceive any difference between similar objects (in the mathematical sense of the word) if we don't have an external thing that retains its size and that we may use as a *tertium comparationis*. The ground of the indiscern-

[46] Saccheri, for instance, who as a philosopher and a theologian might have been interested in the concept of space, was too much of a classical geometer to envisage a new geometry of space. It is quite clear that the *Euclides vindicatus* (1733) is still an essay in the geometry of figures, and regards the problem of proving the Parallel Postulate as a question concerning the systematization of axiomatics, and not the metaphysical nature of space. Lambert's *Theorie der Parallellinien* (1766), on the other hand, was already opening toward a consideration of space as a structure and the Parallel Postulate as stating a property of space. The following generations of mathematicians were all "modern" as far the object of geometry is concerned (János Bolyai, for one, wrote a *Scientia spatii*, not an essay *on parallel lines*).

[47] The main text on the topic is the essay *Uniformis locus* from 1692: "Uniformis locus seu sibi congruus, dici potest, cujus partes congruenter terminatae sunt congruentes. At locus sibi similis est, cujus partes similiter terminantes sunt similes. Loca sibi similia sunt non alia quàm recta, planum, spatium ipsum. Loca uniformia sunt omnia loca sibi similia, et praetera alia, nempe ex lineis quidem arcus circuli et helix cylindrica, ex superficies verò sphaerica et cylindrica …" (in De Risi 2007, pp. 582–85).

[48] The name isogeny and one of the most famous attempts in this direction come from Delboeuf's *Prolégomènes philosophiques de la géométrie* (1860).

ibility of different *similar* cosmic configurations is once more to be found in the relational structure of Leibniz' space, that he opposed to that of Newton's followers.[49] We see that the isogeny of space (its flatness) should follow from the very definition of it as a relational *ordre des situations*, as is required by Leibniz' conception of a geometrical demonstration of the Parallel Postulate. Should we push the argument further, we might see how the distinction between quality and quantity (that metaphysical distinction on which Wallis had grounded his proof of the Postulate) is fundamental in Leibniz' epistemology and theory of perception; and if we consider space itself as nothing more than a representational form of our sensibility, we see that Leibniz could not but endorse Wallis' point of view, and in fact he could have claimed the truth of the Parallel Postulate on purely philosophical grounds.[50] Leibniz, however, never produced an explicit metaphysical proof of it. It should be remembered that his attempts to prove the Postulate did not aim to establish its truth (which he had never doubted), but to devise a sound geometrical demonstration of it through his new *analysis situs*.[51]

Philosophy, however, crept into the attempts to prove the Parallel Postulate in another and unexpected way: through the Principle of Reason. It is well known that this principle was widely employed by Leibniz as a most powerful weapon in metaphysical debates, and in particular in his philosophical war against the Newtonian "idol" of an absolute space. It was through his great Principle of Determinant Reason, that he claimed to have wiped out absolute space, void, and atoms, like leaves in the wind (*ut folia vento*).[52]

In fact, the transformation of the Postulate into a feature of space, and its enrollment among the properties of "uniformity" in a broad sense (isotropy, completeness, connectedness, homogeneity, and so on), pushed Leibniz in his late years toward

[49] Leibniz' discussions on the possibility of shrinking or enlarging the universe in proportion are clearly connected with his better-known objections about the inanity (and impossibility) of moving the whole material world to another part of space, that he raised against Clarke and Newton; this would be *agendo nihil agere* (*Fourth paper to Clarke*, § 13; in Robinet 1957, p. 89). On the reduced universe, see for instance Leibniz' late *Replicatio ad Stahlianas Observationes*, xxi, § 13 (Carvallo 2004, p. 134); the claim from his young years that "si toutes les choses du monde qui nous regardent, estoient diminuées en même proportion, il est manifeste, que pas un ne pourroit remarquer le changement" (A ii, 1, n. 158, p. 380; A iii, 2, n. 79, p. 227–28; the complete passage is quoted in the following *note* 48 in § 4); or a passage in the *Specimen geometriae luciferae*: "si Deus universum orbem cum omnibus partibus proportione eadem servata redderet majorem, nullum esset principium id notandi" (GM vii, p. 266).

[50] I have to refer again to my De Risi 2007 for Leibniz' phenomenological theory of similarity and its connection with the entire Leibnizian theory of knowledge. I only mention a couple of passages (from the geometrical writings) in which Leibniz does not simply claim that his perceptual definition of similarity is used to recognize and distinguish quality and quantity, but also that these important philosophical notions are in fact themselves first defined by this phenomenological act: see *Specimen geometriae luciferae*, GM vii, p. 276; *Initia rerum mathematicarum metaphysica*, GM vii, pp. 18–19.

[51] On Leibniz' aloof reaction to Wallis, and his possible involvement with Borelli's theory of similarity, see above *notes* 29 and 30 in § 2. Wallis' principle about the possibility of similar but unequal figures is stated in Leibniz' *text 14* below, dealing with the theory of parallels.

[52] Leibniz to George Cheyne, September 25th, 1716: "Hoc uno axiomate, spatium reale, vacuum, atomi, difflantur ut folia vento" (still unpublished in L.Br. 154, Bl. 2r).

the idea that it might be itself connected with the Principle of Reason. In the essay *In Euclidis* πρῶτα (1712), Leibniz confessed his incapability to prove the Postulate starting from the simple definition of parallels (or space) and the law of identity, and devised a new demonstration that fundamentally relies on an application of the Principle of Reason. The proof is based, roughly speaking, on the idea that there is no *a priori* reason why space should have a specific curvature (be it positive or negative), and thus it has to be Euclidean. A lack of symmetry, in other words, should always be justified; and in the *simple definition* of absolute space as an order of situations one cannot find any reason for a non-vanishing curvature.[53]

The main problem with this line of reasoning (that will be repeated again and again in the history of the attempts to ground the necessity of Euclidean space even in times in which the consistency of non-Euclidean geometries was already well-established) is that flatness is regarded as a kind of uniformity, and in the end as a negative property: a mere lack of determination. In effect, if we look at Leibniz' general construction of an *analysis situs* as a geometry of space, we have to praise precisely the fact that he was able to go beyond the idea that space was just an amorphous background devoid of structural properties. This new awareness of geometrical space was grounded in Leibniz' exact definitions of uniformity, homogeneity, continuity, and so on, as *positive* properties of a structure, rather than of simple ungeometrical features of an undifferentiated extension (such as the old notion of a primary matter). It is true, however, that Leibniz wavered on the topic, and even the notion of continuity (so important for him) was sometimes again regarded as a simple consequence of a lack of determination.[54] This is even more true for flatness, which was only partially recognized by Leibniz as an independent property of space. It was very easy for him to relapse into the old conception of uniformity as a non-property and thus claim (in our terms) that a vanishing curvature (space's positive property of having a Euclidian metric) is just the lack of any curvature whatsoever ("uniformity" as a merely negative feature); so that in the absence of any reason to give a specific curvature to space, it will be flat. Leibniz' metaphysics of space, which regarded this latter as an ideal order to be determined *a priori*, prevented him from embracing the solution of many later thinkers (from Gauss to Lobachevsky, from Riemann to Helmholtz), who envisaged an empirical determination of a spatial feature (curvature) that could not be attained by logic alone. The Principle of Reason becomes then the tool to provide a determination that could not be grounded, neither in the analysis of the concept of space nor in the experience of the world.

---

[53] For a modern discussion on the employment of the Principle of Sufficient Reason in mathematics, in connection with the symmetries of space (and the theory of groups), see Birkhoff 1940.

[54] This is not just a regressive feature of Leibniz' geometry, but it is also deeply grounded in his epistemology, where the notion of continuity is connected with the confused perception of extension (as in reality every part of *physical* space is actually divided to infinity). Continuity is thus a kind of ideal property that only arises through perceptual underdetermination: a purely negative feature of physical extension. Nonetheless, Leibniz was able to positively characterize it as a property of ideal, geometrical space (see above *note* 36). The physical (merely negative) characterization, however, sometimes creeps into the mathematical one, and in any case the two are not perfectly coordinated.

The main epistemological point, however, is that Leibniz in this text seems to think that the Principle of Contradiction alone is not enough to prove the Postulate. In fact, the Principle of Reason appears not to be reducible to that of Contradiction, and thus guarantees a real extension of the deductive powers of a system.[55] The problem, however, is that Leibniz regularly claimed that the Principle of Reason should only apply to contingent truths and physical features of the natural world; it should be the governing principle for the actual world, while that of Contradiction rules possible worlds (and the actual one among them, of course), as well as the ideal entities.[56] Leibniz' main and recurrent example of the application of the Principle

---

[55] There is just one place, a remark in the margins of a book by Stegmann, in which Leibniz seems to claim that the Principle of Reason may be reduced to that of Contradiction (see the transcription in Jolley 1975). Given the complete isolation of this statement in Leibniz' *corpus*, however, most interpreters believe that it cannot be considered to outweigh that overwhelming majority of discussions in which Leibniz assumes the irreducibility of the "two great principles of our reasoning". Note that Leibniz claims that a perfect geometrical demonstration (grounded in the complete analysis of the notion involved) may only have one realization: there are not different perfect demonstrations of the same claim. Here a quote from a writing of the same year of the *In Euclidis* πρῶτα: "Diversae demonstrationes ejusdem theorematis, si ad perfectam analysin deducantur, debent postremo redire ad idem. Alioqui nondum habemus perfectam analysin. Cum quaedam axiomata sunt assumta sine demonstratione et eorum conclusio quaedam bis demonstrari potest, modis qui in se invicem non resolvuntur, fieri potest ut collatione diversarum demonstrationum ipsum assumtum demonstretur" (*Demonstratio omnimoda*, in De Risi 2007, p. 603). This seems to imply that a proof through the Principle of Reason (which is itself *a principle*, not a provable statement) cannot be reshaped into a proof without it.

[56] See for instance the *Second paper* to Clarke (almost contemporary to the *In Euclidis* πρῶτα): "Le grand fondement des Mathematiques est le Principe de la Contradiction, ou de l'Identité, c'est à dire, qu'une Enontiation ne sauroit etre vraye et fausse en même temps, et qu'ainsi A est A, et ne sauroit etre non A. Et ce seul principe suffit pour demonstrer toute l'Arithmetique et toute la Geometrie, c'est à dire tous les Principes Mathematiques. Mais pour passer de la Mathematique à la Physique, il faut encor un autre Principe, comme j'ay remarqué dans ma Theodicée, c'est le Principe du besoin d'une Raison suffisante …" (Robinet 1957, pp. 35–36). See also the *De principiis praecipue contradictionis et rationis sufficientis*: "Adeo ut quicquid non mathematicae necessitatis est (quemadmodum formae Logicae et veritates numerorum), id omnino hinc [the Principle of Reason] sit petendum" (A vi, 4a, n. 170, p. 806); or again (in connection with Archimedes), the *Specimen inventorum*: "Hoc principio non indiget Arithmetica et Geometria, sed indiget Physica et Mechanica, eoque usus est Archimedes" (A vi, 4b, n. 312, p. 1616). As a matter of fact, Leibniz elsewhere connects the Principle of Reason to *every* kind of truth, saying that it is just a reformulation of the containment definition of truth: if a statement is true, then there must be a logical reason for that, and it is that the predicate is contained in the subject. See for instance the essay on *Primae veritates* (A vi, 4b, n. 324, p. 1645). One should distinguish, however, this purely logical form of the Principle from the one that Leibniz employs in studying the laws of nature (or the Parallel Postulate), which is a much stronger statement dealing with symmetries and uniformities of the relevant object. A connection between the two meanings is to be found in the *Introductio ad encyclopaediam arcanam*, in connection with the possibility of proving geometrical axioms as well as Archimedes' statical principle: "Et quidem nihil omnino fit sine aliqua ratione, seu nulla est propositio praeter identicas, in qua connexio inter praedicatum et subjectum non possit distincte explicari, nam in identicis praedicatum et subjectum coincidunt aut in idem redeunt. Et hoc est inter prima principia omnis ratiocinationis humanae, et post principium contradictionis, maximum habet usum in omnibus scientiis. Ita axiomata Euclidis, *si aequalibus addas aequalia* etc. sunt corollaria tantum hujus principii, nulla enim redid potest ratio diversitatis. Similiter axioma quo utitur Archimedes initio sui tractatus *De aequiponderantibus*, hujus principii

of Reason, in fact, is the justification of Archimedes' postulate in statics that equal weights hung on a balance with equal arms are in equilibrium.[57] It is hard to see how this claim might be reconciled with the use of the Principle of Reason in a purely geometrical proof. In fact, Leibniz' demonstration of the Parallel Postulate in the *In Euclidis* πρῶτα makes no appeal to any physical property of the existing world, nor to the concept of perfection (the best possible world), nor anything else besides purely mathematical terms.

We have already mentioned that Leibniz' basic conception of truth is the containment of the predicate in the subject of a proposition. It was on this notion of the analyticity of truth that he intended to ground his definition of the alethic modalities. If the analysis of the subject and the predicate shows their identity (or containment) in a finite number of steps, the truth in question is necessary; otherwise it is contingent. For instance, the angular sum of two right angles is contained in the notion of a triangle, and can be deduced from it in a finite number of steps (the geometrical proof of *Elements* I, 32, possibly reformulated in Leibniz' new *analysis situs*); this property is thus necessary for the triangle. The fact that Caesar crosses the Rubicon is also contained in the complete notion of Caesar (otherwise it would be false that Caesar crossed the Rubicon), but there is no way to deduce Caesar's action from the concept of Caesar in a finite number of steps; it is true, but contingent, that Caesar crossed the river.[58] This doesn't mean, however, that we cannot know contingent truths except by direct

---

nostri (nihil est sine ratione) corollarium est. Quoniam vero nobis non est datum semper omnium rationes a priori reperire, hinc cogimur fidere sensibus et autoritatibus et maxime et perceptionibus intimis et perceptionibus variis inter se conspirantibus" (A vi, 4a, n. 126, pp. 529–30).

[57] The Archimedean postulate is the first in *De aequilibris planorum* (or *De aequiponderantibus*, in the main Early Modern editions), in *Opera*, vol. 2, p. 124 (ed. Heiberg). Leibniz' discussion of it is recurrent; see for instance the *Principium scientiae humanae*: "Ex his sequitur Axioma maximi Usus ex quo pleraque in re physica et morali derivantur: Nihil evenire cujus ratio reddi non possit, cur sic potius quam aliter contigerit. Exempli causa inter totius Staticae fundamenta ponitur ab Archimede duo pondera aequalia *A* et *B* et aequaliter a centro motus, *C*, distantia in aequilibrio esse, quod corollarium est hujus axiomatis nostri, nam si qua eveniret diversitas, utique aliqua reddi potest diversitatis ratio (per nostrum axioma) quod non potest fieri (ex hypothesi) cum omnia sese utrinque eodem modo habere ponantur; itaque nec diversa inde consequi possunt" (A vi, 4a, n. 157, p. 671). Cf. *De principiis* (A vi, 4a, n. 170, p. 806); *Principia logico-metaphysica* (A vi, 4b, n. 324, p. 1645); *Second paper to Clarke* (Robinet 1957, p. 36). Note, however, that it is a half-mathematical and half-physical statement, and even if Leibniz explicitly states that it obviously pertains to the *res physica*, it is difficult to claim that it is properly contingent: to the point that Leibniz claims elsewhere that its truth is known *a priori*: "Il faut aussi remarquer qu'on ne peut prouver un effet naturel par les seuls principes intellectuels, si ce n'est lorsque tout est égal de part et d'autre; car en ce cas l'experience n'est pas necessaire: comme, cette demande d'Archimede, *les poids égaux en distances égales, pesent également*, peut passer pour un principe intellectuel; car où prendroit-on l'inégalité, et d' où pourroit-elle proceder, puisque tout est pareil de part et d'autre? Mais cette autre demande, *les poids égaux en distances inégales, pesent inégalement*, a besoin d'experience" (*Remarks to Mariotte*, in A vi, 4b, n. 222, p. 1040). The same is true in a passage from *Sur ce qui passe les senses et la matiere*, where Archimedes' principle is grounding the *a priori* science of mechanics (GM vi, pp. 489–90). The example from statics is also relevant for the Parallel Postulate itself (see below, § 5.4).

[58] For this famous Leibnizian definition of necessity and contingency, see for instance the *De natura veritatis* (A vi, 4b, n. 303, p. 1515–18); or § 61 of the *Generales Inquisitiones* (A vi, 4a, n. 165,

experience; we may find other principles that govern the behavior of things and guide us in the realms in which analysis is impossible. The Principle of Contradiction (in the form of the identical axioms) is the only principle that governs analysis. The Principle of Reason is then employed as a further help (an extra deductive tool) to understand and relate contingent truths among themselves, in short to be able to have a science of contingent laws of nature.

Upon this logical conception of the modalities, however, Leibniz often superimposed more classical notions of contingency and necessity, as well as a few characterizations of them through a possible-world semantics. Should we simply equate necessity with truth in all possible worlds, the two collapsing concepts of necessity will imply the deductive completeness of geometry: a geometrical truth is valid in all possible worlds (as space is the same in all of them), thus it is necessary, thus it is deducible in a finite number of steps from the definitions of the concepts involved (and eventually from the definition of space). This will leave no room for the need of the Principle of Reason in a geometrical proof. It is true, however, that Leibniz does not need to endorse a complete equivalence among his different notions of necessity, and he may think that even though a finitary deducible proposition is true in all possible worlds, the inverse implication is false, and there are propositions true in all possible worlds that are however not deducible from definitions and the Principle of Contradiction alone (in a finite number of steps).[59] If this is the case, the Parallel Postulate, proved *a priori* through the Principle of Reason and applying to all possible worlds, may still be contingent in the logical sense, and thus unprovable from the simple definition of space (by the law of identity). It is not obvious, however, that this may be the solution to our epistemological problem. We may expect that a proposition true in all possible worlds, but still contingent, should be a physical statement, possibly *a posteriori* or at least based on some experience, and in any case dependent on the features of the substances (monads) that ultimately ground every possible world; while a proposition bearing on an *ideal* entity, like space, should probably be entirely founded in the "realm of eternal truths" just like any other geometrical statement.[60] Another solution here may be to relax the de-

p. 758); the *Origo Veritatum contingentium* (A VI, 4B, n. 327, pp. 1659–64); the *De Libertate* (A VI, 4B, n. 326, pp. 1653–59); and the *Theodicy* (GP VI, p. 414).

[59] In his remarkable *De natura veritatis* (1686) Leibniz explicitly says that finitarily deducible propositions are true in all possible worlds: "Hinc jam discimus alias esse propositiones quae pertinent ad Essentias, alias vero quae ad Existentias rerum; Essentiales nimirum sunt quae ex resolutione Terminorum possunt demonstrari; quae scilicet sunt necessariae, sive virtualiter identicae; quarumque adeo oppositum est impossibile sive virtualiter contradictorium. Et hae sunt aeternae veritates, nec tantum obtinebunt, dum stabit Mundus, sed etiam obtinuissent, si Deus alia ratione Mundum creasset" (A VI, 4B, n. 303, p. 1517). I am not aware, however, of any instance of the inverse claim among Leibniz' writings.

[60] Several natural laws, and in particular the principle of inertia and some forms of the principle of continuity in physical change, are sometimes regarded by Leibniz as contingent and sometimes as necessary, even though one cannot tell what kind of necessity this might be. The same is true for the all-important example of the balance that we mentioned above. I gave some more examples of this peculiar necessity in my DE RISI 2007. In a deeper consideration, one should say (see above, *note* 43) that space itself is constructed by the perceptual faculty of a monad (as it is just an ideal

mands of the Leibnizian conception of truth as containment of the predicate in the subject (which was in any case very basic) and claim that there are necessary truths which are not analytical.[61] This would lead us to claim that the Parallel Postulate, as it is proved in a purely conceptual way and without any reference to the (real or possible) physical world, but still through the Principle of Reason, is a *synthetic a priori* statement. It is clear, however, that Leibniz' epistemology is unprepared to accommodate this kind of proposition.

As Leibniz' proof of the Parallel Postulate through the Principle of Reason is only attempted in one essay, and remained without comments or epistemological elaborations, I don't think it is useful to belabor the point any further. It was probably only an experiment in his geometrical studies, like many others the great man conceived in his private and very private notes. It is however remarkable that Leibniz' very last attempt to prove the Parallel Postulate ended up with the implicit admission that non-Euclidean geometry is impossible and false as a description of space, but nevertheless *consistent*, as its negation does not imply a contradiction but just a violation of the Principle of Reason.

"form of intuition"), and it is thus reasonable to claim that the properties of space depend not just on the definition of it as a total situational order, but also on the features of sensibility in general (which, in turn, depends on the inner nature of the monads). In this sense, the "deduction" of an Euclidean metric based on the phenomenology of perception (rather than on a theory of parallels) that is hinted at in the *Initia rerum mathematicarum metaphysica* (more or less in the same years of the *In Euclidis* πρῶτα), may go in the same direction of a spatial feature which is not provable through a conceptual analysis but only by stronger (but still universal and *a priori*) assumptions. Finally, a suggestive but rather anachronistic presentation of Leibniz' views would be to take the Principle of Reason as a sort of infinitary inference rule (like the so-called ω-rule), and claim that the Parallel Postulate is a true but unprovable statement, and Euclidean geometry an incomplete system (with respect to syllogistic inferences and those other finitary rules that Leibniz might have accepted as purely logical).

[61] Note that we have a similar problem with the Principle of Reason itself, which (if it is undeducible from that of Contradiction) cannot possibly be considered an analytical truth.

Leibniz' Attempts to Prove the Parallel Postulate

4

4.1 Preliminary Definitions.

After this sketch of Leibniz' epistemological views, we are now able to briefly discuss the properly mathematical texts that Leibniz devoted to the theory of parallels and the proof of the Parallel Postulate. Given the epistemological relevance of the system of (real) definitions in Leibniz' geometry, however, it may be of some use to mention the most important characterizations that Leibniz employed in defining the basic geometrical terms.

First of all, the definition of a straight line. We already know that a good definition of a straight line should dispense, in Leibniz' view, with the axioms relative to it (such as that between two points there is one line, or that two lines cannot have a common segment), as the properties expressed by the axioms should be enclosed in the definition itself. The search for a viable characterization of "straightness" had been one of the central foundational concerns since Antiquity, and it was often discussed in connection with the Parallel Postulate. Several modern geometers claimed, in fact, that the lack of a good definition of a straight line was also the main reason why the Parallel Postulate could not be proved with ease.[1] This view is

[1] The idea is widespread. Saccheri said that the nature of the straight line must be investigated more profoundly; for, without such efforts, geometry could scarcely grow beyond its current infancy (*Euclides vindicatus*, pp. 34–35); Lambert shared similar ideas: cf. *Theorie der Parallellinien*, § 10, p. 149. I may also mention D'Alembert's *Éclaircissemens sur différens endroits des Élémens de Philosophie,* XI, where he discusses the definition of parallel lines and the difficulty in proving the Postulate, drawing on his article *parallèle* for the *Encyclopedie* (see below, *note* 98), and adds that the real reason for the failure is the missing definition of a straight line: "On parviendroit peut-être plus facilement à la [a theory of parallel lines] trouver, si on avoit une bonne definition de la ligne droite; par malheur cette definition nous manque. … La definition & les proprieties de la ligne droite, ainsi que des lignes paralleles, sont donc l'écueil, & pour ainsi dire, le scandale des élemens de Géometrie" (*Mélanges*, vol. 5, pp. 202 and 206–207). The same was later maintained by Legendre: "C'est sans doute à l'imperfection du langage vulgaire et à la difficulté de donner une bonne definition de la ligne droite, qu'il faut attribuer le peu de succès qu'ont obtenu jusqu'ici

V. De Risi, *Leibniz on the Parallel Postulate and the Foundations of Geometry*,
Science Networks. Historical Studies, DOI 10.1007/978-3-319-19863-7_4

quite reasonable as, even though the Postulate is not (properly speaking) a principle *about* straight lines, it is nevertheless concerned with straight lines. Consider, for instance, that the main feature of a non-Euclidean theory of parallelism (in hyperbolic geometry) is the existence of approaching straight lines that do not meet (asymptotic parallels); if one could disprove this possibility, the Parallel Postulate would be established; but an asymptotic behavior of lines seems to depend on their curved and convex "shape" (as the usual Euclidean asymptotic lines are, obviously, not straight); a good definition of straightness (one might hope) would have dispelled the possibility of *asymptotic straight lines*, thus opening the way to a full-fledged demonstration of the Postulate. This point of view, moreover, seems to be especially in agreement with Leibniz' views, as a good definition of a straight line should dispense with several axioms, and thus (maybe) with the Parallel Postulate as well.

Leibniz' writings on *analysis situs*, in fact, present many attempts toward a definition of straight lines. Leibniz' definitions are very numerous, and it is useless to try to give a complete catalogue of them now, but they can roughly be arranged into three categories.

First, there is the definition of a straight line as the shortest line between two points. This definition gained several supporters in the seventeenth century, as it seems to be the only mathematical definition with a clear meaning and a straightforward formulation; it was employed, for instance, by Borelli, Giordano, Fabri and Hobbes.[2] Leibniz appreciated it, too, but preferred to comply with the Archimedean tradition of taking this property of a straight line as a *characterization* of it more than a definition proper.[3] The reason is obviously that the notion of "straightness" seems not to be captured by "shortness", and the original intent of a definition of a straight line is that of expressing the clearly intuitive idea of a lack of convexity or concavity (to use Clavius' words). Leibniz explicitly mentions the fact that the property of shortness is never used in any proof of the *Elements*, and it is not useful to solve any

les géomètres …" (*Réflexions sur différentes manières de démontrer la théorie des parallèles*, 1833, p. 372). This was also the opinion of Kästner, who had such a great influence on the German researches on the Parallel Postulate that eventually resulted in the creation of non-Euclidean geometries: "Der Grund, warum man in diesem Axiome [the Parallel Postulate] nicht die Evidenz der übrigen findet, ist … daß man von der geraden Linie nur einen klaren Begriff hat, nicht einen deutlichen" (*Ueber den mathematischen Begriff des Raums*, 1790; p. 414). But one could find Greek sources to the same effect, as Aristotle's statement (*Phys.* B 9, 200a2-4) that from the essence of the straight line it follows that the sum of the angles of a triangle is equal to two right-angles.

[2] See BORELLI, *Euclides restitutus*, p. 4 (ed. 1658; with a criticism of Euclid and Clavius); GIORDANO, *Euclide restituto*, p. 5; FABRI, *Synopsis geometrica*, p. 27; HOBBES, *De corpore*, XIV, 1 (in *Opera*, vol. 1, pp. 153–55).

[3] As we will see commenting *text 9*, Leibniz was able to grasp the structural likeness between a straight line in a plane and a circle on a spherical surface, thanks to their minimizing properties. He even devised several papers devoted to the analytical determination of the shortest line on any curved surface whatsoever, a problem that was proposed by Johann Bernoulli in 1697 (see a couple of texts in DE RISI 2007, pp. 592–95, and a brief discussion of the debate at p. 88 n. 97). It is not strange, then, that he could not think of assuming the Archimedean characterization of a straight line as a proper definition.

foundational question.[4] He had a discussion with Giordano on this point, in which he appreciated that Giordano had been able to employ the definition of a straight line as the shortest line to implement his proof of the Parallel Postulate: which, being concerned with the notion of *equidistance*, might rely as well on the idea of the shortest distance between two lines. This was an improvement, in Leibniz's eyes, on Borelli's treatment of the Postulate.[5] In the end, however, Leibniz would have preferred that Giordano had employed in his foundational efforts a definition of a straight line that Leibniz showed he knew elsewhere: that of a rotational axis.[6]

---

[4] We have already seen (see above, *note* 9 in § 3) that Leibniz thought that Euclid's defective definition of a straight line forced him to assume several axioms to characterize it; these axioms (and not the definition) are employed in demonstrations. But also Archimedes' definition (which is in fact a characterization) of the straight line, cannot properly be used in the proofs of the *Elements*. Leibniz claims, on the contrary, that a definition that takes into consideration the shape of the line is needed "ad Euclideas demonstrationes perficiendas" (*In Euclidis* πρῶτα; GM v, p. 185). See also Leibniz' letter to Schenk from the same year 1712: "La definition de la Ligne droite, que je viens de donner, paroit estre la plus propre à faire le supplement des Elemens. Celle qui a plû à Archimede ne paroist pas avoir plû à Euclide, autrement il n'auroit point entrepris de prouver que dans un triangle rectiligne un coste est tousjours moindre que les deux autres, ce qui n'est qu'un corollaire de la definition ou proprieté d'Archimede" (in De Risi 2007, p. 621). The idea, however, recurs since Leibniz' first studies in geometry: "Sic Euclides nuspiam utitur definitione lineae rectae in suis demonstrationibus, sed ejus loco adhibuit quaedam pro axiomatis assumta; at Archimedes cum longius vellet progredi, coactus est ipsam lineam rectam resolvere, eamque definire, inter duo puncta minimam" (*Elementa calculi*, from April 1679; in A vi, 4a, n. 57, p. 197).

[5] Borelli defines the straight line as the shortest one, but then he proves *Elements* I, 21 (that a side of a triangle is shorter than the sum of the other two) without making recourse to his definition; moreover, in justifying his Axiom 14 (the Parallel Postulate), he appeals to Clavius' arguments (which Borelli does not count as real proofs of it) which in turn rely on Clavius' definition of a straight line. Leibniz criticized Borelli on this point, and in his copy of *Euclides restitutus*, he commented on Borelli's definition of a straight line: "Non utitur autor hac definitione, cum uti potest, v.g. in axiomate 14 (quod assumit post prop. 13), utitur potius Clavij definitione seu explicatione, ut axioma confirmet. Et posita hac definitione non erat usu demonstraret propositionem 21 quod duo trianguli latera sint majora tertio" (remark to *Euclides restitutus*, ed. 1658, p. 5; a similar remark also appears in Leibniz' marginal note to Proposition 21, p. 47).

[6] Here is Leibniz' first letter to Giordano, in which he praises the fact that Giordano was able to employ the Archimedean definition of a straight line in his proof of the Parallel Postulate: "Sed in universum in horum axiomatum de recta demonstrationibus illam difficultatem reperio, quod in eas nullo modo ingreditur definitio rectae nec ulla rectae proprietas axiomate aliquo praemittendo contenta. Definitio enim rectae a te assumta est quod sit brevissima inter duo puncta, quae pulchre uteris pro parallelarum proprietate, sed hic eam non adhibes nec aliud de recta axioma praemittis. Itaque in omnibus istis demonstrationibus posset alia quaecunque linea pro recta assumi quod tamen male fieret. Itaque videtur aliquid his demonstrationibus deesse" (A ii, 2, n. 71, p. 295). Giordano answered that he had only used the Archimedean definition because it is easier to grasp ("At Heronis definitione sum usus, non alia de causa, nisi quia visa mihi est accommodatior Tyronum intellectui"; A ii, 2, n. 72, p. 296), but he considered the Euclidean one as the most correct, and the principle of every other characterization. To which, Leibniz replied that no one was able to employ Euclid's definition in a proof: "Interim quacunque demum utamur notione rectae, eam influere ut ita dicam, oportet, in theoremata quae de recta demonstrare volumus, alioqui ignotum est, utrum ea quae demonstramus ad eam rem pertineant, cujus data est definitio. Idque in illis demonstrationibus Euclideorum Axiomatum quas a Proclo et Clavio mutuatus es desiderare me jam innui, etsi hoc in responsione tua praeterieris. Quomodo enim ex iis sciemus pertinere ad lineam brevissimam

The definition of the straight line as the locus of points which remain fixed in the motion of a body with two fixed points, i. e. the axis of rotation, is in fact the second and most important definition of a straight line that Leibniz uses in his writings on *analysis situs*. It appears in a number of different formulations, and it is even formalized in the new symbolism of the *characteristica geometrica*. Its clearest mathematical characterization is that the straight line is the only line between two points that is unique under isometries (that is, any other line connecting those two points has at least another isometric realization – imagine the reflection of that line in a plane). It is also connected with Leibniz' notion of *uniformity*, as the straight line so defined has the property that every segment of it is congruent to any other segment of the same length, by a spatial isometry (its curvature being null). This definition of a straight line was sketched in Antiquity (by Heron and Proclus), and was taken up again in the seventeenth century by Roberval, Giordano and others.[7] Its importance for Leibniz is enormous, as it is by far his preferred definition, and the one that he normally employs in his foundational attempts in elementary geometry, including those aimed at proving the Parallel Postulate. We should remark that a similar definition was to be used by Saccheri, Gauss and Lobachevsky.[8]

The third definition of a straight line that Leibniz uses is that of a self-similar curve. Modeled by analogy with the above-mentioned definition, in which each segment of a line is congruent with any other of equal length, this definition says

inter sua puncta extrema. Caeterum cum propositum esset in *Euclide* tuo omnia qua licet exacte demonstrare, fortasse non diffiteberis rectius suppleri casus qui ad perfectionem demonstrationis desiderantur quod tironibus opinor praejudicium facere non poterat" (A II, 2, n. 73, p. 303).

[7] The ancient definition is just that of a line that remains fixed if its extremities are fixed (Heron, *Definitiones* 4; Proclus, *In Euclidis* 110), and the explicit example by Heron is that of a line reflected in a plane (not a rotational axis in space). Already al-Haytham, however, interpreted the definition as a rotational axis in space (see Sude 1974). Among the moderns, Roberval accepted this definition (*Elemens*, p. 100), and employed a Postulate to guarantee that (1) a body can rotate, and (2) the locus of points fixed in the rotation is a line (see Jullien 1993 and Hayashi 1998). Leibniz obviously appreciated these foundational concerns, but wanted to prove these kinds of postulates (we saw some Leibnizian quotes in this direction in *note* 42 in § 3). Giordano meant to employ such definition in his *Archimede* (a follow up of his *Euclide*, that was never published), even though he was not satisfied with it: "In meo Archimede sic rectam lineam definivi: la linea revoluta intorno ai suoi estremi immoti, le di cui parti ritengono sempre il medesimo sito di prima, la chiamo recta linea, sed fateor: ea in definitione non acquiesco: expungam ipsam, et Euclideam, quam optimam duco (atque rectitudinem explicat) reponam" (Letter to Leibniz; in A II, 2, n. 71, p. 297). Giordano's foundational perplexities were grounded in the fact that this definition of a straight line requires the notion of a plane or a body (which is rotated), and he thought that higher-dimensional objects should be defined starting from lower-dimensional ones, and not vice versa (otherwise it would be *ponere (ut aiunt) currum ante boves*; p. 296). Leibniz was not worried by this criticism, as he was devising new, structural definitions of extension and dimensionality that need not define dimensions in a recursive way. He added, however, that the rotating body has to be convex (*Uniformis locus*, in De Risi 2007, p. 583).

[8] See Saccheri, *Euclides vindicatus*, Lemma 1 to Proposition 33 (pp. 70–72), where he employed this definition to prove that two straight lines do not enclose a space (and several other results, such as that the diameter bisects the circle). On Gauss' definition, see Zormbala 1996. As for Lobachevsky, see his *New principles of geometry*, ch. 2, § 25, as well as a similar definition in *Pangeometry* (p. 280 of the French edition); cf. Lechalas 1912.

that the straight line is the only line any segment of which is *similar* (in the proper geometrical sense) to any other segment (of any length) of the same line.[9] This definition is also widely employed in Leibniz' foundational attempts, and plays a role in his proofs of the Parallel Postulate. Given that the very concept of similarity is in question with the Postulate, one could say that the truth of the Parallel Postulate is the condition for such a definition or rather that this last notion of a straight line, taken as a *real* definition, already implies the truth of the Parallel Postulate.

Leibniz, in fact, aspires to prove the *reality* of these definitions of a straight line, that is, the intrinsic possibility of their objects. Moreover, he would like to prove that all these definitions are in fact equivalent, and describe the same object. We will not enter now into the details of Leibniz' proofs of existence and equivalence, but we should remark that the proofs meant to establish the reality of the definition of a straight line through rotation (the second definition we mentioned) require a proof of the axiom stating the uniqueness of the line between two points, and (as we have noted) the reality of the definition of a straight line through similarity requires a proof of the Parallel Postulate. In the end, thus, Leibniz' attempts to prove the Postulate may be led back to the problem of finding (and justifying) a good definition of a straight line.[10] On the other hand, since Leibniz is not fond of the definition of a straight line as the shortest line, and his definition of a self-similar line already implies the truth of the Postulate (which has to be preliminarily proved in order to allow the definition), the general conclusion is that in his attempts to prove the Parallel Postulate Leibniz was mostly to employ a concept of straight line as an axis of rotation. This definition of a straight line is thus the starting point for his demonstrations.[11]

Besides the definition of a straight line, we are obviously concerned with Leibniz' definitions of parallel lines. Leibniz' usual preference for the definition of parallels as equidistant lines, however, makes the (real) definition of parallels more an aim than a starting point in the proofs of the Parallel Postulate (as such a definition already "contains" it implicitly). We have already remarked that there are several epistemological reasons why Leibniz prefers this definition of parallel lines, the most important of which is precisely that it allows us to deduce the Postulate, thus

[9] Any homothety of a factor λ applied to a curve transforms the curvature of it by a factor $1/\lambda$. If we want a similarity (that is, a homothety and a rigid motion) to bring a segment of a curve to coincide with the whole given curve (also bounded or closed), we have $\lambda > 1$, and thus the curvature lessening at every point. The maximal curvature of the given curve, thus, may only be reached by the transformed segment if it is zero, that is, if the curve is a straight line.

[10] In De Risi 2007 (pp. 240–64), I have followed the thread of Leibniz' theorems of characterization of a straight line to introduce the most relevant results of his foundational analysis, like the proof of the Parallel Postulate.

[11] Note that Leibniz normally thinks that a definition of a straight line should be good enough to prove the axioms that two straight lines cannot have a segment in common or enclose any space, but he rarely mentions the Parallel Postulate as a possible consequence of such definition alone: to this effect we also need a definition of parallel lines. See for instance §§ 21–23 of *Analysis geometrica propria* from 1698 (in GM v, p. 176), in which Leibniz writes about the straight line and its axioms; cf. with the above-mentioned § 2, on the definition of parallels, in *In Euclidis* πρῶτα (GM v, p. 201; *text 36* below), where he says that Euclid was forced to assume the Parallel Postulate since he lacked a good definition of parallels (not of straight lines).

shaping the principles of geometry as a system of definitions without (non-identical) axioms. Moreover, there are clear historical facts behind his choice, as almost all the mathematicians working on the foundations of geometry in the seventeenth century accepted (uncritically or not) the definition of parallels as equidistant lines. A further reason for Leibniz' preference doubtless lies in the fact that the notion of equidistance is easily generalized as applying to curves and, while the concept of non-incident curves in a plane seems to be useless in geometrical practice, that of equidistant curves may find several applications. Hobbes insisted on this matter, and Leibniz followed him in this from his very first days as a geometer. A more general definition of parallelism, that could be applied to both straight lines and curves, is thus one of the aims of Leibniz' studies on the Parallel Postulate. Lastly, and importantly, we should insist that Leibniz' concern with the principles of geometry is somehow a late interest of his (one that arose when he was still young, but already an accomplished mathematician and analyst) and thus he thought that the notion of non-incident straight lines was quite useless in higher mathematics. If we look at the occurrences in Leibniz' mathematical writings (that is, those occasions on which he uses the notion without trying to justify it or to prove the Parallel Postulate), we immediately see that their mutual equidistance is the only relevant feature in theorems and proofs. One of these contexts is the foundational debate on the Calculus, which often passes through a discussion of the *indivisibles* employed in previous decades (by Kepler, Cavalieri, or Wallis), which are parallel lines or surfaces at an infinitesimal distance from one another. The concept of (infinitesimal) equidistance is here the key to proving or disproving the most relevant theorems on the matter, and Leibniz was to insist on the notion of equidistant parallel lines in this connection. Another field of enquiry in which parallel lines frequently pop up is geometrical optics, where parallel light rays are reflected or refracted into converging or diverging rays; in this context, the incidence properties are not rich enough to allow a thorough mathematical analysis, and one needs the effective calculation of angles and distances, and thus takes parallel lines to be equidistant (at a given distance) much more than non-incident. Even richer is the use of equidistant curves in connection with the Calculus, and several studies by Leibniz deal with the generation and analytic properties of evolutes and involutes to given curves (a concept widely studied by Huygens when Leibniz was in Paris); not to mention the obvious employment of equidistant curves in geography and cosmography (the circular parallels on a sphere being the most evident example), which Leibniz deals with at length in his physical and cosmological writings.

In general, we can say that the incidence structure of space (or its affine structure) was not yet grasped as an interesting geometrical concept, and the notions of intersecting or non-intersecting lines were considered too weak to be usefully employed in mathematics – in an age that was celebrating the highpoint of the classical theory of measure through the Calculus.[12] It should be recalled, moreover, that the basic, un-

[12] We have already remarked that, before Leibniz, a few geometers did, in fact, employ the notion of situation in mathematics. We don't have, however, any treatment of the Parallel Postulate by Desargues or Pascal (we may guess that Pascal's ideas on the subject were expressed in Arnauld's

defined, concept of Leibniz' entire endeavor in *analysis situs* is the notion of distance. If to reach an *essential* definition of parallel lines (following Leibniz' epistemology) we have to resolve this concept into its first and simple conceptual marks, such as the concept of *situs* itself (which is equated to distance), it plainly follows that the notion of equidistance will be the main feature of the theory of parallelism in the *analysis situs*. So, even though Leibniz' geometry of *space* was a considerable step forward toward the construction of a geometry of positional relations that could complement the classical geometry of measure and magnitudes; and even though Leibniz' work on the foundations of geometry showed, from time to time, purely non-metrical concerns (as for instance his studies on the Axiom of Pasch[13]); nevertheless the *analysis situs* remained mainly a reshaping of classical, metrical geometry.

In sum, Leibniz' epistemological program in the foundations of geometry, as well as his expertise as a working mathematician in other fields, pushed him to accept a notion of parallelism that resorts to equidistance.

Since we have mentioned Leibniz' studies on the Calculus, it should be remarked that one of the aims of the *analysis situs* and the inspection of the foundations of geometry was surely that of a sounder justification for the new infinitistic procedures. These kinds of foundational reflections are clearly connected with the theory of parallels, as the tangent spaces (as we would call them today) employed in classical analysis for the linearization of functions are in fact affine and Euclidean structures. I am not aware of any direct connection made by Leibniz himself (or other mathematicians in his age, truth be told) between the Parallel Postulate and the notions of differential or derivative. Leibniz' peculiar path to the foundations of the Calculus, however, passes through the possibility of the transformation by *similarity* of the "differential triangle" (the linearized right triangle whose sides are the infinitesimal abscissa and ordinate of the given function); and the possibility

---

Nouveaux elemens, but we have no clear clue on the matter). Patrizi's *Nuova geometria* shows a vast construction dealing with the incidence properties of straight lines, based on the notion of situation (Books VI–XIV). However, Patrizi misses this opportunity, and defines parallel lines as equidistant lines (re-introducing a metrical notion), thus "proving" the Parallel Postulate (in Book XIV) with a usual paralogism.

[13] It is highly remarkable that Pasch's Axiom (in the form now usually called the "crossbar theorem") was in fact discussed in Leibniz' work on *analysis situs*. See a passage from an untitled essay from 1679: "*Recta ab uno trianguli angulo usque ad basin ducta, quae latus non est, tota intra triangulum cadit.* Nam quia ab angulo educitur punctum habet cum quolibet angulum facientium commune, jam et punctum habet cum basi commune, ergo cum quolibet Trianguli latere punctum aliquod commune habet. Ergo inter angulum & basin triangulo egredi non potest, nam ambitus eius id est aliquod laterum alicubi secabit, et ita cum aliquo laterum duo puncta cummunia haberet. (Superest, quod potest extra triangulum tendere ad basin ita ut triangulum circumeat, et hoc fieri non posse demonstrandum)" (LH XXXV, I, 2, Bl. 14r; transcribed in Echeverría 1980, vol. 2, pp. 78–79). Before Leibniz, Roberval had explicitly assumed a similar statement as Postulate 2 of the First Book of his *Elemens*: "En tout triangle BCD auquel le point F pris dans la base, aura servi de point fixe pour décrire la superficie du triangle; si par dedans le triangle, sur la même superficie, on mène la droite FG, prolongée suffisamment, elle coupera l'un ou l'autre des côtés CB ou CD en un point. Soit ce point G, ou même, elle coupera tous ces deux côtés au point C qui en ce cas conviendra au point G" (p. 139). A similar principle was also to be found in Claude Richard's 1645 axiomatization of the *Elements*, as Axiom 29 (see Richard, *Euclidis*, pp. 17–18).

of this sort of transformation, as we know, is equivalent to the assumption of the Parallel Postulate.[14]

There is a sense, thus, in which the Leibnizian Calculus needs the foundational efforts in the theory of parallels. We should note, moreover, that the similarity (or rather, homogeneity) between finite and "infinitesimal" quantities should not necessarily imply that the latter are actual infinitesimals, as Leibniz may think of them as quantities smaller than any *given* quantity, in the modern sense.[15] This issue has some relevance, since if we accept the existence of actual infinitesimals and thus disregard Archimedes' Axiom, several deductions about the theory of parallels are excluded. For instance, the Parallel Postulate is no longer equivalent to the possibility of transformations through similarity (Wallis' proof *essentially* employs the Archimedean Axiom), nor to the fact that a (Saccheri's) quadrilateral has four right angles. Even though Leibniz could not know about these results (which were proved by Hilbert, Dehn, and others in the twentieth century), it was worth mentioning the issue. Following some interpretations of Leibniz' notion of the infinitesimals, in fact, one may claim that the general environment of his mathematics is non-Archimedean.[16]

---

[14] The best presentation of the meaning and use of the differential triangle in Leibniz' foundational studies on the Calculus is Bos 1975. The paper also contrasts Robinson's non-standard analysis with Leibniz' own project.

[15] Even though it is quite clear that Leibniz embraced at some point a conception of an infinitesimal as an arbitrarily small *finite* quantity, the interpreters are however divided on the details and the real extent of the endorsement. Leibniz' clearest stance of a finitist conception is probably his *Quadratura arithmetica circuli*, which was published and commented in Knobloch 1993 (see also Knobloch 2002b), and is now the main essay in A vii, 6, n. 51. Note also that Leibniz' perceptual definition of similarity, which is so important and recurrent in his geometrical writings, does not allow one to think of an actual infinitesimal as similar to a finite quantity (since the infinitesimal cannot be perceived). Even more properly, an infinitesimal is rectilinear (as it is the linearization of the function) while the curve is not; thus, Leibniz said that the one is just *homogeneous* to the other (not *similar* in the proper sense), where homogeneity means that one can be transformed into the other through a bi-continuous transformation (see for instance the *Specimen geometriae luciferae*, GM vii, pp. 282–84). Leibniz, moreover, wanted to equate in some sense his own conception of homogeneity to that of Euclid (*Elements* V, def. 3 and 4), where homogeneous magnitudes are just Archimedean magnitudes (see his letter to Marchetti, in A iii, 7 n. 196, p. 780). It seems thus that Leibniz could not think (at least in these geometrical foundational contexts) about actual ("non-standard") infinitesimals. In a letter to L'Hospital, however, he considers infinitesimals as non-Archimedean quantities, again in reference to Euclid's theory of proportions (A iii, 6, n. 135, pp. 416–17). Pascal held similar views about indivisibles: see Gardies 1981.

[16] It may be noted, in any case, that already Nasīr ad-Dīn explicitly stated (without proof) the need of the Archimedean Axiom for the applicability of his researches on the quadrilaterals to the proof of the Parallel Postulate; and Wallis' proof about similarity also depended on Nasīr ad-Dīn. On the Archimedean Axiom and Wallis' proof, see Cassina 1958. On the interior angle sum of a quadrilateral, see Dehn 1900; as well as Hilbert 1899. Note that Hilbert's famous calculus of segments in this book is aimed at building a *purely geometrical* theory of similarity, which is also Leibniz' main goal in his studies on the topic. Hilbert's point, however, is properly to skip the Euclidean (Eudoxian) use of the Archimedean Axiom.

4.2 Leibniz' Youth and the Parisian Years (*Texts 1–4*).

Leibniz' serious engagement with foundational issues in elementary geometry does not date back before 1679, when he was already a well-trained and well-known mathematician and had made the most relevant discoveries that opened the way to the Calculus. It is worthwhile, however, to consider some texts on parallel lines from his earlier years, to better understand some later developments.

We know that Leibniz first studied elementary geometry when he was a student in Leipzig in 1661, but apparently without foundational concerns. Later on, he read extensively on "modern geometry", from Cavalieri to Descartes and Barrow (and Hobbes), but without referring any longer to Euclid's *Elements*.[17] On the other hand, it seems that Leibniz began to conceive his epistemological ideas on mathematics already in the mid-1660s. While his thoughts on the matter underwent important developments during his life, the basics of his conception (which we have sketched above) seem to be already settled well before any thorough reading of elementary geometry. In particular, the young Leibniz was convinced that every axiom of geometry can be proved starting from definitions and identical propositions, and he seems to have already envisaged a logical proof that the whole is bigger than the part. He was aware, in any case, that Clavius had attempted to prove some of the axioms.[18]

During his mathematical training in Paris (1672–1676), Leibniz came across some editions of Euclid, and began to learn elementary geometry from an advanced standpoint. His attention, however, was drawn toward the new results and the higher mathematics of the age, and it seems that he did not dedicate much time or intellectual resources to truly familiarizing himself with the *Elements*. Still in 1676 he wrote to Foucher that he had only been able to read Euclid in the way in which "*on a coustume de lire les histoires*".[19]

---

[17] On Leibniz' apprenticeship as a mathematician, see Hofmann 1974 and Taton 1978.

[18] We have a statement by Leibniz himself to this effect, who in the late *Historia et origo calculi differentialis* (1714) says: "Et puer adhuc logicam versans animadverterat ultimam veritatum a ratione pendentium analysin abire in haec duo: definitiones, et veritates identicas, solas necessariarum vere primitivas indemonstrabilesque; et cum objiceretur ipsi, veritates identicas inutiles et nugatorias esse, ipse contrarium etiam experimentis ostendebat, atque inter alia jam tum monstrabat Axioma illud magnum, Totum esse majus parte, demonstrari per syllogismum, cujus major propositio esset definitio, minor esset propositio identica" (GM v, p. 395). We might want to doubt Leibniz' own reconstruction of his youthful thoughts, and for instance Couturat's claims that the idea of employing identical propositions as axioms came to Leibniz some years later (as there is no trace of it in the *Preface to Nizolius* of 1670, in which Leibniz only mentions definitions as principles of demonstration; cf. Couturat 1901, p. 186 n. 1). In any case, Leibniz "proved" the mereological axiom already in 1671–1672 (*Demonstratio propositionum primarum*, in A vi, 2, n. 57, p. 482–83; cf. also A ii, 1, n. 109, p. 355); and we are sure that he had already fully developed his new epistemology of definitions and identical axioms just a few years later, during his stay in Paris (see for instance A vi, 3, n. 22, p. 328; but also the text from 1676 quoted above in *note* 20 in § 3). In the same *Demonstratio*, we find Leibniz' mention of Clavius: "Notum est quaedam Axiomata ab Euclide assumta, a Clavio aliisque in Theorematum numerum redacta esse, quasi paulo difficiliora" (A vi, 2, n. 57, p. 480).

[19] Here the complete quote from the letter to Foucher: "J'aimois tousjours des livres qui contenoient quelques belles pensées, mais qu'on pouvoit parcourir sans s'arrester, car ils excitoient en moy des

We know, however, that he studied Fabri's *Synopsis geometrica* in mid-1673 and annotated it with very many marginal notes; he even composed a separate essay on Fabri's book, giving it no less a title than *Characteristica geometrica*. This last composition shows a Leibniz who is still a beginner in geometry; he is clearly concerned with some foundational problems, but his reflections remain quite unrefined and mostly have a philosophical or metaphysical flavor that discloses a weak grasp of the truly important issues in elementary geometry. Among these notes, we find Leibniz' first reference to parallels (*text 1*), which merely repeats Fabri's definition of them as equidistant lines. This was apparently the first exposure Leibniz had to the definition of parallel lines that he endorsed throughout his life.[20]

Among Leibniz' studies in elementary mathematics, we may also mention a fragment *De secandis parallelis* (*text 2*) dealing with a simple result in Euclidean geometry; this is almost an exercise, proving a simple consequence of Thales' Theorem (*Elements* VI, 2). The Parallel Postulate is here assumed as self-evident (in the sentence: "*patet* aequales esse AD, BL, CM", which is only true in Euclidean geometry) and not discussed further. Leibniz' interest in this specific geometrical result, however, may be of some interest as it may be considered one of the foundations of Euclidean statics. Saccheri was to discuss the very same theorem, generalizing it to hyperbolic and elliptic geometry, and the grounding of a non-Euclidean mechanics was to employ this generalization. It does not appear that Leibniz was aware of the significance of this theorem during his youth; Thales' Theorem was to surface again in later studies, however, and the relation between statics and the theory of parallels was to be an important issue in connection with the Principle of Reason.[21]

Working on both higher mathematics and a new metaphysics, in 1676 Leibniz wrote several essays on infinity, some of them touching on the concept of parallelism, which (as many mathematicians observed with concern) is entangled with the idea of an infinite extension of straight lines. Even though these texts do not discuss the Parallel Postulate or the definition of parallels, they have some relevance in showing a few lines of reasoning that will surface again in later studies. In the fragment *Linea infinita est immobilis* (*text 3A*), Leibniz tries to prove that an infinite straight

idées, que je suivois à ma fantasie et que je poussois où bon me sembloit. Cela m'a encor empeché de lire avec soin les livres de Geometrie, et j'ose bien avouer que je n'ay pas encor pu gagner sur moy de lire Euclide autrement qu'on a coustume de lire les histoires" (A II, 1, n. 120, p. 247). The idea is repeated in the *Historia et origo calculi differentialis*: "... et Euclidem vix satis attente adspexerat, aliis plane studiis intentus" (GM V, p. 398). We have to point out, however, that in an essay *De constructione* from 1674, Leibniz boasts of having read Euclid *very attentively*, contrary to the manner in which he was used to proceeding with other books: "Ego cum Euclides elementa, nuper attente legerem, quod fateor a me fieri perraro, aut potius si de integro libro quaestio sit, facto hactenus nunquam" (A VI, 3, n. 45, p. 420).

[20] On Fabri's reading, see again the *Historia et origo* (GM V, p. 399).

[21] Saccheri proves a version of this theorem as Proposition 20 of *Euclides vindicatus* (p. 28). Similar results were used by Archimedes (in *De aequilibriis planorum* A, Prop. 9–10 and 13–14; in *Opera*, ed. Heiberg, vol. 2, pp. 140–58) to find the barycenter of a triangle and thus grounding classical Euclidean statics. See below, § 5.4. We may also remark that this elementary result will be used by Leibniz in his essay on the quadrature of the circle: cf. Proposition 1 of *De quadratura arithmetica circuli* (A VII, 6, n. 20, pp. 178–79; A VII, 6, n. 51, p. 522).

line cannot rotate, otherwise at some point in its rotation it would become parallel to another (fixed) given line, so that an infinite segment of it should pass from below to above the given line "at the same time" (however small the angle of rotation); and this is impossible.[22] We see here a philosophical construal of the notions of limit and infinity, which has no real connection with the Parallel Postulate, but underlines well some of the difficulties and paralogisms that other mathematicians would fall prey to, especially when they were confronted with the possibility of asymptotic straight lines. This is also the kind of difficulty that Borelli and other similarly-oriented geometers might have envisaged arising from the Euclidean definition of parallels as non-intersecting lines.[23] In later texts of the same year, Leibniz claims that an infinite line cannot be moved along its own direction, otherwise the infinite ray generated at point A would be translated into the infinite ray generated at point B (on the same infinite straight line), and thus an infinite line (the ray at A) would coincide with a *smaller or larger* infinite line (the ray at B). Leibniz is clearly tackling the usual difficulties with infinite measure and cardinality, and in fact he notes himself that a similar translation would occur also in the set of natural numbers (passing from n to $n+1$), so that (1, 2, 3, …) would coincide with (2, 3, 4, …), the part with the whole; and this (in Leibniz' view) is impossible.[24] Leibniz' willingness to retain the Euclidean axiom that the whole is bigger than the part also in infinite sets, leads him to state that every straight line has a well-defined middle point and, consequently, to discuss the possibility of a center of the whole of space.[25] He further claims that a

[22] ARTHUR 2001B remarks that the fragment may have had some metaphysical relevance for Leibniz independently from geometrical considerations, as it constitutes a counterexample to the Cartesian tenet that everything extended is movable, a thesis that Leibniz was discussing with Malebranche in the same months (cf. A II, 1, n. 125, pp. 403–406). The argument is important, of course, because for Leibniz *space itself* is extended but immovable, while Cartesians denied the ontological independence of space from extended matter.

[23] The limiting case between incidence and parallelism was one of the most difficult topics to be mathematically treated before the birth and rigorization of the Calculus. The masterpiece of Saccheri's *Euclides vindicatus* is in fact the complete (and exact) theory of limiting asymptotic lines in hyperbolic geometry, treated with the classical methods of synthetic geometry (Propositions 29–32, pp. 63–70). Still, as is well known, Saccheri himself blunders in the following Proposition 33, stating that two asymptotic lines would have a common segment at infinity, which is not consistent with the definition of a straight line (such a definition is thus, for Saccheri, the logical source of the Parallel Postulate). Several other geometers, less versed than Saccheri in the classical methods, claimed to have proved the Parallel Postulate with kinematic arguments not far from the abovementioned reflections of Leibniz.

[24] Leibniz extends his reasoning to the integers in the essay *Numeri infiniti* (A VI, 3, n. 69, pp. 495–504). Among recent scholars, Richard Arthur has articulated and defended Leibniz' notion of cardinality as an attempt to keep the axiom that the whole is bigger than the part, abandoning in turn other set-theoretical principles that ground Cantor's later construction. See below at *note* 27 for some references.

[25] Leibniz' views on space in the 1670s were not yet developed in the direction of his mature writings, that is toward idealism and structuralism. He was still discussing (if not accepting) theories of extension as a substance, and thus he could play with the idea of a center of cosmic space. Even in these early writings, however, the notion is only accepted dubitatively. See also *De mente, de universo, de Deo* from December 1675, in A VI, 3, n. 57, p. 465.

straight line cannot be moved in its own direction, nor rotated around a point which is not its center, even though he is ready to accept (at least provisionally) that such a line could be translated parallel to itself. This engenders a discussion on the infinite strip of a plane bounded by two parallel lines (*text 3B*), which he will take up again later in the course of a proof of the Parallel Postulate (see below, *text 22*). Here we see, for instance, that there is only one parallel line to a given straight, passing from the center of space (a weak form of Playfair's axiom), or that the line equidistant (i. e. parallel) to a line that exactly bisects a plane is not itself a bisector. Finally, a last fragment of the same year (*text 3C*) seems to point directly to the Parallel Postulate, as it states that from the fact that two unbounded straight lines are not parallel, one can prove that they meet; and the same line of reasoning may be applied to bounded lines which are extended sufficiently to meet. We do not know what Leibniz means here by "non parallel lines", but as he explicitly intends to *prove* their intersection, it is likely that the definition of parallelism will not be that of "not-intersecting lines" but rather that of equidistance.[26]

We won't belabor these texts, as they are more relevant for Leibniz' conception of infinity than for that of parallel lines, and they are too fragmentary to offer a theory of parallelism. In any case, Leibniz would ultimately manage to escape from these kinds of difficulties with infinity and cardinality through his doctrine of *actually infinite syncategorematic wholes*, which admits the possibility of an actually infinite object, but denies that such an object (e. g. a straight line) could ever be considered as a real whole (as *one* object).[27] In the following discussions of parallels, Leibniz (with the sole exception of the above-mentioned *text 22*) was to only consider limited

[26] We may remark that the final sentence of the fragment, that "straight lines, unbounded in themselves, are bounded by us or by bodies" seems to entail that Postulate One ("to draw a straight line from any point to any point") should be read for Leibniz not as a construction but as an existential assumption ("between two points there is a straight line"), as it happens, for instance, in Hilbert's axiomatics. This would not be a complete novelty, since already Francesco Patrizi, discussing his new metaphysics of space as an extension independent from matter, had explicitly reinterpreted Euclid's constructive postulates as existential statements. See, for instance: "Lineam enim negamus, a nostra mente, aut arte in infinitum posse produci: attamen eam quae punctis finita est, iis liberata infinitudimem sui natura, fatemur subire" (Patrizi, *De spacio mathematico*, p. 20r; *Nova philosophia*, p. 66v; for a geometrical application, see also *Della nuova geometria*, p. 34). The hint came in fact from Proclus, *In Euclidis*, 86 and 101, who however apparently intended a different meaning. In any case, Leibniz' mature theory of space seems to imply that a line (as a system of situational relations) exists independently of any constructive procedure. The constructive stance, however, is still working in his interpretation of the postulates of classical geometry (cf. *In Euclidis* πρῶτα, GM v, pp. 205–206), and in the theory of constructive definitions which are grounded (however) in the properties of an *existing* situational order.

[27] In late 1676, Leibniz would attempt to solve the difficulties of a continuous motion of infinite wholes through the concept of *transcreatio*: he sees motion as a series of continuous annihilations and re-creations of the same object in different spatial positions (see *Numeri infiniti*, in A vi, 3, n. 69, p. 500; and, most of all, the *Pacidius Philalethi*, in A vi, 3, n. 78, p. 560). His final stance, however, would be the above-mentioned distinction between infinite aggregates (which have no unity themselves) and infinite wholes. This distinction is already suggested as an exit strategy from the problem of cardinality in the text *Numeri infiniti* (A vi, 3, n. 69, p. 501), and would become one of the most important features of Leibniz' mathematical and metaphysical thought. See the ample discussion on the topic in Levey 1998, and Arthur 1998, 1999, 2001A, 2001B, 2013A.

segments, as was the case in classical geometry, and all these speculations on infinity were to be left aside.

Leibniz was effectively employing these kinds of reflections on infinity in his attempts to obtain and ground new mathematical results. In his groundbreaking studies on the arithmetical quadrature of the circle (which was a source of Leibniz' fame as a mathematician, as well as many of his own ideas on the Calculus), for instance, he considers a point of intersection between a straight line and an asymptotic line to be at an infinite distance (*text 4*B).[28] The considered straight line is not the axis of the asymptote, as in this case they would not have any intersection whatsoever, but nevertheless converges "just a bit" with it – converging, say, at an infinitesimal angle. The result is that the straight line is of *infinite length*, but nevertheless *bounded* by the point of intersection with the asymptote; nor is it perfectly parallel with another straight line, but just converging a bit with it. In the years of the invention of the Calculus, it is not hard to find a similar imaginative statement on infinity and intersections. In general, Leibniz would like to distinguish infinite unbounded lines from infinite bounded lines, claiming that the former (as instances of a categorematic infinite) have no common measure at all with a finite segment, while the latter (which are syncategorematically infinite) may be considered as *composed* by an infinite number of finite segments. The analogy is with points, which do not compose a line even if taken in an infinite number (since adding point to point bears no extension), and infinitesimal lines, which added together an infinity of times do make up a finite segment; so that an infinitesimal line is to a finite segment as a finite segment is to an infinite bounded line, while no proportion exists among points, finite segments and unbounded lines.[29] Even though Leibniz learned to better control his mathematical tools with the passing of the years, these kinds of suggestions had a lasting presence in the discussion of asymptotic behavior in the eighteenth century.[30]

Another source for Leibniz' early conceptions of parallelism in relation to infinity is the geometry of conics and the "projective" methods there employed. In particular, we know that Leibniz had access to Pascal's lost treatise on conic sections, where the French mathematician, following Desargues' own reflections on the subject, assumed that two straight lines whatsoever always meet (at finite or infinite distance). Leibniz endorsed Pascal's views and took parallelism as a special case of intersection in a point at infinity (*text 4*A).[31] This point of view is obviously very fruitful in the

---

[28] Cf. also another version of this text in A VII, 6, n. 51, p. 547.

[29] See the relevant passage in the Scholium to Proposition 11 of *De quadratura arithmetica* (A VII, 6, n. 51, pp. 548–49; cf. A VII, 6, n. 20, pp. 212–13).

[30] I am again especially thinking of Saccheri, who in the aforementioned Proposition 33 of his *Euclides vindicatus* was willing to claim the provability of the Parallel Postulate, just because he takes the point of intersection at infinity between a line and an asymptote as behaving like a common point of the plane (cf. also his Corollary 3 to Proposition 26, in which he extends further two asymptotic straight lines *after* their meeting point at infinity; *Euclides vindicatus*, p. 56).

[31] While he was in Paris, on June 4th, 1675, Leibniz received from Perier some manuscripts on conic sections written by Pascal. On August 30th, 1676, he wrote to Perier a letter explaining the content of these manuscripts; he also transcribed from them a chapter of Pascal's unpublished treatise on conics. Since in the following century Pascal's manuscript went lost, Leibniz' very partial transcrip-

general theory of conics and one of the fundamental principles of the geometry of projections. It also has the advantage, in Leibniz' view, of representing the notion of non-incidence as a limiting case of that of incidence, thus allowing a *continuous* transition from the one to the other. In several other papers from his mature years, Leibniz would restate the same idea, that the general principle of continuity (which plays an important heuristic role in his thought) grounds the transit from incidence to parallelism, and that this kind of continuity allows one (for instance) to formulate general properties about the different conic sections (see below *texts 11, 25, 27*). Leibniz wrote several papers on conics and quite a number of essays on the theory of perspective; none of them have been published at present, but they promise a rich geometrical inheritance that will deserve further investigation.[32] It should be remarked, finally, that if Leibniz had these seminal results in projective geometry in mind, which assume that there are no non-intersecting straight lines, it was quite natural for him to define parallelism through the notion of equidistance: since parallel lines are then those kinds of *intersecting* lines that are equidistant at any finite distance, and only meet at infinity – while the Euclidean definition of parallels through non-incidence appears to be devoid of use.

It is obvious, in any case, that in all these writings on parallels and infinity, Leibniz was not attempting to read or ground Euclid's *Elements*, but was pursuing his own programs of a *scientia infiniti* and a general renewal of mathematics. This situation would begin to change in the following years.

4.3 The First Studies on the Parallel Postulate: 1677–1689 (*Texts 5–16*).

By the end of 1676, Leibniz had left Paris and arrived in Hannover. In the following years, far from the center of the mathematical world, he cultivated his own projects toward a new science, and in particular he developed several essays of *characteristica*, both *universalis* and *geometrica*. In the course of these studies, he happened to

tion is the only evidence we have about Pascal's ideas on the topic (besides a short and juvenile *Essai pour les coniques* that Pascal did publish when he was sixteen; cf. Pascal, *Oeuvres*, vol. 1, pp. 252–60). In this unpublished treatise, we find the following statement: "Definitio 3. Duae rectae aut plures quomodocumque sint positae, dicuntur semper concurrere, et quidem ad distantiam vel finitam, si se in eodem puncto intersecent, vel infinitam, si sunt parallelae" (Leibniz, *Briefwechsel mit Mathematikern*, p. 137; A vii, 7, n. 4, pp. 85–91. Leibniz' entire transcription of Pascal's *Generatio conisectionum* is also to be found in Pascal, *Oeuvres*, vol. 2, pp. 234–43). Elsewhere in the same connection, Leibniz himself remarks: "Duae lineae parallelae concurrere intelligantur, etsi locus concursus infinite absit" (Pascal, *Oeuvres*, vol. 2, p. 233). The idea comes from Desargues' *Brouillon project d'une atteinte aux evenemens des rencontres du Cone avec un Plan*, from 1639, which Pascal had read in his youth (now easily available in the commented edition of Field&Gray 1987). A complete description of Pascal's work on conic sections is in Taton 1964; on Leibniz' reading of Pascal's manuscripts, see Mesnard 1978; on Desargues' use of points at infinity, see Field 1997. The notion of parallels as lines meeting at infinity was later employed by D'Alembert in the entry *Parallèle* of the *Encyclopedie* (vol. 11, pp. 905–6).

[32] See for instance the papers in LH XXXV, xi, 17, Bl. 19–20; and LH XXXV, xi, 1, Bl. 1–13.

read Euclid with care, and began to be interested in a foundational work on geometry that strictly followed the text and the principles of the *Elements*.

Already in January 1677 he was reading Kaspar Schott's *Cursus mathematicus* and annotated it in a short essay that he again called *Characteristca geometrica* – the same title as his reading notes on Fabri's *Elements* some years before. Leibniz' notes on Schott, an author not at all versed in foundational issues, show that he was now concerned with Euclid's system of principles and not just with the philosophical discussion about the nature of continuous quantity (as in the notes to Fabri). Coming to the Parallel Postulate, Leibniz showed sincere surprise that anyone could have considered it as an unprovable or evident principle (*text 6*), and even though at the time he did nothing to try to prove it, this observation may well be considered the first engagement of Leibniz with our problem.[33] More or less at the same time, Leibniz was also reading Barrow's edition of Euclid, and he remarked again (*text 5*) on the difficulty of the Parallel Postulate. In this connection, he also added that the Postulate is false if one doesn't assume that the two straight lines are in the same plane; this issue will surface again later (see especially *text 31* below).[34]

The year 1679 is the *annus mirabilis* of Leibniz' work on his new geometry, and he wrote innumerable papers on *characteristica geometrica*. The year begins with a *Demonstratio axiomatum Euclidis*, which however doesn't discuss the Parallel Postulate, and culminates in the draft of his most important essay of *Characteristica geometrica* (from August 20th), some excerpts of which Leibniz sent to Huygens asking for a judgment (he got a slating). The writings of this year, important as they are for the development of Leibniz' geometrical and logical ideas, are more concerned with a general program for proving the axioms than effective attempts in this direction, and mostly deal with the invention of a new mathematical symbolism. We find however a few interesting items.

A passage in one *Scheda* preparing the great *Characteristica geometrica* (*text 7*) shows Leibniz' interest in the Parallel Postulate, and concocts a program to prove it from the definitions of a plane and a straight line. Even though this preliminary sketch will not find its way into the final version of the *Characteristica*, and therefore will not enjoy the wider and refined treatment reserved for the other geo-

[33] Note that Echeverría, who first published Leibniz' *Characteristica geometrica* from 1677, mistakenly claims that this paper is composed of reading notes to the *Elements* in the version of Dechales. Dechales' *Cursus mathematicus*, however, employs a system of principles that doesn't match with Leibniz' notes, while Schott's own *Cursus* is easily recognizable as the background of the *Characteristica*, as it presents to the reader an overabundant list of principles (4 postulates and 26 axioms) that finds few equals in other seventeenth-century textbooks of geometry. Schott, on the other hand, took all of them as provable. On Schott's *Cursus*, see Knobloch 2011.

[34] It is difficult to exactly date Leibniz' marginal notes to Barrow. It should be noted, however, that Barrow's edition was republished in 1676 (the first edition was from 1655), and already in May 1678 Leibniz refers to it and to his personal remarks on it: see A ii, 1, n. 177, p. 622. Moreover, among the *marginalia* we find the following remark: "Si corpus quodcunque movetur duobus punctis immotis, habetur simul generatio rectae et circuli. Nam puncta omnia quiescentia cadent in *rectam*, mota describent *circulum*" (p. 7 of Leibniz' copy of Barrow). The latter passage is quite similar to Leibniz' fragment *Generatio quidem rectae et circuli* (see above, *note* 42 in § 3), which was dated by Echeverría to 1676.

metrical topics touched in the longer essay, it is not difficult to guess what Leibniz had in mind. The 1679 *Characteristica* devises a combinatorial approach to the intersections of planes and lines, which can be easily generalized into an abstract theory about the incidence properties of affine subspaces of any dimension. This theory has some points of contact with Grassmann's *Ausdehnungslehre*, and has been, in fact, considered an "anticipation" of it.[35] Leibniz' interest in the matter, in any case, is aimed at giving exact algebraic proofs about the dimension of the subspaces cut by the intersection of other subspaces (e. g. that two planes intersect in a line; a line and a plane in a point; and so forth). The combinatorial rules that Leibniz gives to this effect are correct, but already assume (embedded in their algorithm) the affine, and thus Euclidean, structure of space. In other words, the Parallel Postulate is taken for granted, and the new algebraic symbolism of the *Characteristica geometrica* simply reflects the Euclidean environment for which it has been devised. It is thus perfectly obvious that Leibniz is able to prove statements equivalent to the Parallel Postulate simply by implementing his formal system. In later years (*texts 20*, *26*, *38*) he was to employ his *characteristica* explicitly to prove the Parallel Postulate. Note, however, that in the present fragment Leibniz also hints that the *uniformity* of a straight line can serve as the ground for the truth of the Parallel Postulate; this remark is surely dependent on Clavius, and was to recur in subsequent writings.

We find among Leibniz papers from 1679 a fragment (*text 8*) in which Leibniz defines parallel lines as equidistant lines and then states a *memorabile theorema* which is a formulation of the Parallel Postulate (*Elements* I, 29, in fact). He doesn't give a proof of it, but it isn't hard to devise, as the proposition expresses a provable and basic property of equidistant lines. Leibniz then notes, and it is a *mirabile paradoxum*, that the line equidistant to a conic section is not itself a conic section. Leibniz was to repeat the same (correct) statement about conics several times (see below *texts 14*, *16*, *21* and *30*), and this was to later become for him the standard counterexample to those who claim that it is obvious that the line equidistant to a straight line is straight. In another text of the same year, he carries out the algebraic calculations required to show that the curve equidistant to a parabola is not itself a parabola.[36] Leibniz would like then to distinguish between *regular figures*, such as the straight line and the circle, whose equidistant line is of the same species (itself straight or circular), and paradoxical lines, such as the (other) conic sections, which

---

[35] Grassmann himself claimed in his essay *Geometrische Analyse geknüpft an die von Leibniz erfundene geometrische Charakteristik*, from 1847, that his own *Ausdehnungslehre* was in fact an improvement on Leibniz' *characteristica geometrica*. This opinion became widespread in the nineteenth century, and was shared by many mathematicians and philosophers (Peano and Cassirer among them). Even though Leibniz' project was in fact quite different from Grassmann's, it is true that Leibniz' rules for the intersections of affine subspaces have something in common with those in modern vector calculus (first stated in complete generality in Grassmann's theory of extension). On the topic, see Heath 1917, Echeverría 1979, De Risi 2007 (pp. 111–13 and 215–22).

[36] This paper is a single page in LH XXXV, xii, 1, Bl. 218, simply titled "Parallelae. Parallela parabolae non est parabola". I do not transcribe it in our collection of texts, since Leibniz' algebraic manipulations (a student's exercise) offer nothing relevant in the foundational domain.

don't have this property. Leibniz realized, of course, that he needed a proof that the straight line is regular (in this particular sense of the word), and that on this proof depends that of the Parallel Postulate.

In this connection, it may be worth mentioning that the fact that the equidistant line to a circle is itself a circle was a further reason for confusion among the mathematicians attempting to prove the Parallel Postulate, since it was regarded, in fact, as an example speaking in favor of the claim that the equidistant to a straight line is itself straight (no one mentioned the conic sections).[37] The relevance of this instance is that Leibniz clearly had in mind that there is a correspondence of some sort between straight lines in the plane and great circles on a spherical surface. The great circles are the shortest and the most determined lines between two points on the sphere, and (moreover) they are uniform lines.[38] The equidistant line to a great circle in a sphere is a parallel, which is no longer a great circle. This is thus the clearest example of an equidistant line to a geodesic which is not a geodesic – an example that Leibniz could clearly envisage. Leibniz still saw the sphere as a surface embedded in three-dimensional Euclidean space. It happens, then, that a parallel in an embedded sphere is itself a circle (if not a *great* circle), and Leibniz was able to think that the equidistant line to a circle is itself a circle, even on a spherical surface (a regular figure, in his terminology). Thanks to this confusion between embedded and absolute structures, and the lack of the concept of a mathematical model, Leibniz missed the most promising example of a non-Euclidean geometry on the spherical surface, and did not properly grasp the full meaning of a geodesic. When he refers to circles as regular figures, he is still working with the classical geometry of figures, not with his new *analysis situs* about space.

In the same year of 1679, in any case, Leibniz attempted to prove that the equidistant line to a straight line is straight, while commenting on Clavius' edition of the *Elements* and the Jesuit's proof of the Postulate (in the last part of *text 9*). Leibniz' short discussion on the topic only restates that the straightness of the equidistant line should be proved (what Clavius just assumes, Leibniz says),[39] and he puts forward an

[37] Clavius' Lemma 1 to his proof of the Parallel Postulate has a long discussion about the fact that the line equidistant to a circle is itself a circle: "Idem prorsus in linea circulari contingit. Nam enim linea inflexa circularem lineam ambiens, cuius Omnia puncta aequaliter à circulari distant, circularis quoque est" (*Euclidis*, pp. 50–51). Wallis used the same example not to bolster the notion of equidistance, but to say that Postulate Three, allowing the construction of circles with any radius, also allowed for the construction of similar figures (*De postulato quinto*, in *Opera*, vol. 2, p. 676). See Saccheri's comments and criticisms on the example of the circle in Clavius (*Euclides vindicatus*, Scholium 2 after Proposition 21; pp. 36–37), and Wallis (*Euclides vindicatus*, Scholium 3 after Proposition 21; pp. 39–40). Leibniz himself commented on Clavius' circle argument with a marginal note to his copy of *Euclidis*, stating: "non semper lineae aequidistantes ejusdem sunt naturae; aequidistans parabolae parabola non est" (p. 90).

[38] As for the shortest line, see (among others) a text from 1695: "In superficie data (e. g. spheroidis) à puncto dato ad punctum datum ducere omnium possibilium in superficie minimam. In plano est recta, in sphaera est circulus maximus" (in De Risi 2007, p. 592). Cf. also *Nouveaux Essais*, II, xiii, § 3.

[39] Since the above-mentioned marginal notes to Clavius' edition were probably written by Leibniz in 1679, I should mention now that on Leibniz' copy of Clavius' *Euclidis* (p. 89) we find a hand-

argument by symmetry. If we take a segment perpendicular to a straight line and move it parallel to itself (as in Clavius' proof), the flow of the free end of the segment describes an equidistant line (correct). The segment, moreover, is also perpendicular to the equidistant line that it describes (also correct, it is the above-mentioned *memorabile theorema*). Now, we may consider the same segment as moving along the equidistant line, tracing with the flow of its other end a line which is equidistant to the equidistant line (obviously correct, since equidistance is a symmetric relation),[40] and moreover straight (by hypothesis). Thus *by symmetry* (but Leibniz does not use this word),[41] the equidistant line, which is generated in the same way as the original straight line, is straight (wrong, it simply doesn't follow).[42] Even though Leibniz' implicit appeal to symmetry in this context is still a bit rough, and does not find support on other, higher principles (e. g. the Principle of Reason), we may already note that it has some points of contact with the further discussion in the *In Euclidis* πρῶτα, (see below, *text 36*).

The same essay also offers another, longer, attempt to prove the Parallel Postulate, which is not connected with the notion of equidistance (the first part of *text 9*). This is more the sketch of a proof than a real demonstration, and Leibniz himself realizes that he needs something additional to be able to draw the desired conclusion. He discusses the case of two lines, AB and CD, that are intersected by a transversal AF as to form a right and an acute angle.

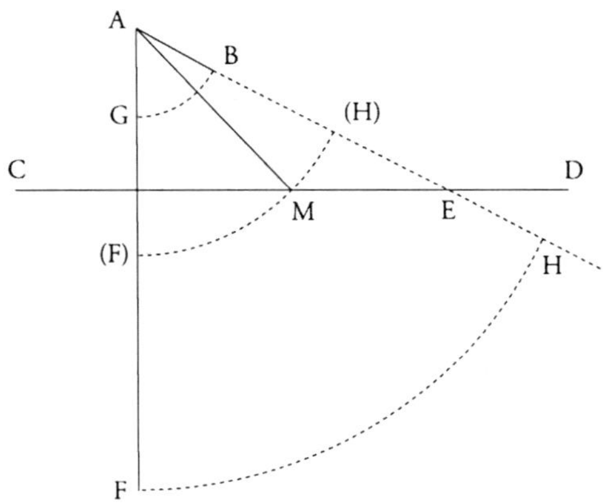

The Parallel Postulate states that, if extended far enough, the two lines will meet. Leibniz imagines the rotation around A of the transversal AF toward the position of the inclined line AB, and considers that the point F will be under the line AC when (at some point of the rotational movement) the line AF reaches the position of AB.

written remark to Clavius' proof about the straightness of the equidistant line, stating that when the Jesuit assumes that the line produced by the flow of the segment is not going up or down, he is already assuming what he would like to prove – straightness ("illud sursum vel deorsum supponit jam rectam").

[40] We may remember that the reflexivity, symmetry and transitivity of the relation of equidistance are often discussed in the Early Modern attempts to prove the Parallel Postulate, and in particular by Leibniz. Here the assumption on symmetry is correct, although unproved. Cf. *note* 19 in § 2.

[41] Following Hon&Goldstein 2008, it seems that Legendre (in 1794) was the first to employ the term "symmetry" in a mathematical context to mean a peculiar *relation* between figures, and not just a single figure with a particular disposition of parts (one symmetrical figure). Leibniz also needs a symmetric *relation* to conclude his proof, as well as to define his *structural* concepts of uniformity or homogeneity of space, etc. But he does not yet have the word.

[42] In hyperbolic geometry, the curve equidistant to a straight line (i. e. an hypercycle) is not straight, and the equidistant to a hypercycle is sometimes another hypercycle and sometimes a straight line (i. e. a geodesic).

Since F is below CD and A is above it, by a continuity assumption (which we needn't discuss now), the extended AB (that is, the rotated AF) meets CD at some point, proving the Postulate. If the length AF is such that when AF reaches the position of AB the point F is already above the line CD (as in the case of A(F) in the diagram) nothing can be concluded, of course. But then, we can take a point F on AF that is far enough from A to do the job. The same will be true for any finite length of AF. In the case in which AB and CD are parallel, however, the same reasoning cannot apply, since the intersection would occur at infinity, and AF would be a rotating unbounded line meeting the unbounded line CD at its extreme (which is impossible, since AF and CD have no ends). Leibniz takes this as a further proof that infinite lines cannot be moved (see *text 3* above). Thus, Leibniz concludes, the present proof by rotation shows that AB and CD intersect, except in the case in which the intersection should occur at infinity; and this is the case in which AB and CD are parallel, or form equal angles with a transversal. Even if we accept the whole reasoning on the immovability of an infinite line (and Leibniz himself observes that it would require more work), the last passage is unwarranted. In hyperbolic geometry, the transversal line AF, rotating, first intersects CD, than it becomes asymptotic with it, and finally it reaches the line AB (assuming, I mean, that AB is parallel to CD in the hyperbolic plane, and thus the Parallel Postulate is false in this instance). That is to say, Leibniz doesn't prove the central statement needed to establish the Parallel Postulate, that the intersection of the rotating AF with CD only occurs at infinity when the angular sum with the transversal is two right angles. Leibniz himself realized that this part of the proof was missing, and abandoned it.[43]

The themes exploited in 1679 return again in a series of texts from 1685. We find another long commentary on Clavius (*text 12*), in which Leibniz attempts to prove several axioms, stating again his grand logicist program in geometry. He then complains about Proclus' and Clavius' proofs of the Parallel Postulate, but cannot offer anything better. He is content to rely on the parallel motion of a segment to produce the equidistant straight line (as Clavius did), and only slightly simplifies the old Jesuit's construction. The text is also relevant because in it Leibniz presses doubts about the possibility of squares, and correctly sees that a proof of the Parallel Postulate is mandatory to have a *real* definition of this figure.[44] Another fragment from 1685 (*text 13*) shows again Leibniz' confidence in Clavius' kinematic procedure to engender equidistant lines which are also straight (or at least mutually congruent, which amounts to the same thing). Leibniz tries to generalize the procedure through the parallel motion of planes and solid bodies, whose flow traces entire families of equidistant lines.

A longer paper from the same year (*text 14*) returns to the issue of the conic sections, and discusses another definition of parallelism. In this paper, Leibniz would like to say that a conic section similar to another conic section may be said to be parallel to

---

[43] We may remark that a proof along the lines of Leibniz' attempt was envisaged by a Medieval Hebrew mathematician (Moïse Ha-Lévi), who may have been relying on Ancient sources: see Lévy 1992.

[44] We may remind that the possibility of squares was already doubted by Borelli in this connection (see above, *note* 30 in § 2).

it. He has already proved, however, that similar conic sections may not be equidistant, since a curve similar to a conic section is a conic section, whereas a curve equidistant to a conic section is not (in general) a conic section.[45] He thus changes the notion of parallelism, and says that two curves are parallel if a straight line cuts both of them at the same angle. Leibniz would like to assume this (quite sketchy) definition as a general notion of parallelism, and states that the true concept of it relies on the idea of *inclination*. This definition of parallelism is relevant since it has several points of contact with Borelli's, and gestures toward a "direction theory" of parallelism, which however Leibniz would not endorse in the future. He did not develop this line of research, and soon came back to a theory of equidistance; the notion of direction, however, continues to pop up, from time to time, in the following years.[46] In any case, the connection between parallelism and similarity is quite important in Leibniz' writings, and in this text he explicitly states Wallis' axiom: "any figure can be contracted or expanded into another bigger or smaller figure which is similar in all respects".[47]

The notion of similarity marks another trend in the studies of the 1680s about the Parallel Postulate. A short essay from 1680 (*text 10*) begins with a statement of Leibniz' recent (1677) definition of similarity through coperception: two figures are similar if I can distinguish them only by perceiving them together, while if I imagine that I perceive them separately, they will be undistinguishable for me.[48] This notion

---

[45] The definition of similar conic sections and the basic theorems about their similarity are one of the main themes of Apollonius' Sixth Book of *Conica*, that (see below *note* 61) was published by Borelli in Leibniz' early years. In it, Apollonius proved that conic sections of different kind (ellipses, parabolas, hyperbolas) are not similar to one another (Propositions 14–15), while all parabolas are similar among themselves (Proposition 11), and states the conditions under which two ellipses or two hyperbolas are similar (Propositions 12–13). A remark by Leibniz that goes in the same direction of *text 14* is to be found already in 1673–1674; see LH XXXV, XII, 1, Bl. 221: "Curvae *B*. et *D*. similes certo modo. Alia est similitudo Apollonii in *Conicis*, ubi videndum an sint verbi gratia parabolae similes et similiter positae, quae in eodem plano secante conum descriptae in plano baseos circulos concentricos proficiunt". This remark is to be published in the next volume of the Academy Edition of Leibniz' works (A VII, 7, n. 15, p. 145), and I am grateful to Siegmund Probst, who communicated it to me.

[46] This concept of inclination may be found in Schott's theory of parallelism (see above), as well as in Mercator's edition of the *Elements* (see below, *note* 93). While an author like Mercator, however, would like to assume the concept of inclination as a primitive notion, Leibniz characterizes it through the angles that a straight line makes with an intersected curve. This is a more classical statement, which reminds of Euclid's own characterization of parallelism (through the Parallel Postulate itself and *Elements* I, 27–29) and was later fully exploited by Borelli. For this reason, I would not say that Leibniz is here properly embracing a "direction theory" of parallelism.

[47] Note that the main efforts in *text 14* are aimed to prove several theorems about the measure of similar figures, that Leibniz considers an important test for establishing a correct notion of similarity: see below, *note* 67.

[48] The perceptual definition of similarity is first given in a letter to Gallois from 1677, which is worth quoting extensively: "Apres avoir bien cherché, j'ay trouvé que deux choses sont parfaitement semblables, lorsqu'on ne les sçauroit discerner que per compraesentiam, par example, deux cercles inegaux de même matiere ne se sçauroient discerner qu'en les voyant ensemble, car alors on voit bien que l'un est plus grand que l'autre. Vous me direz: je mesureray aujourdhuy l'un, demain l'autre; et ainsi je les discerneray bien sans les avoir ensemble. Je dis que c'est encor les discerner non per memoriam, sed per compraesentiam: parce que vous avez la mesure du premier presente,

of similarity will be retained by Leibniz during his whole life, and he used it, in fact, to build a bridge between his theory of perception and geometry: so that in the end, his idealistic metaphysics of space as a "form of intuition" (rather than a substance or an attribute of a substance) rests on this very notion of a perceptual similarity. The problem is, of course, to actually reconnect the perceptual definition with the original definition of Euclid, and thus be able to employ the perceptual similarity in geometry: to this effect, Leibniz produced several papers, spanning from the 1670s to the end of his life.[49] The present fragment is one of these, and after stating the perceptual definition, it goes into several other definitions of basic geometrical terms through similarity, the most important of which is that of a straight line as a self-similar line.[50] Then it tackles the Euclidean definition of similarity, that similar figures have equal angles and proportional sides. As Leibniz here *takes for granted* that his perceptual definition of similarity applies to figures with proportional sides, he would like to *prove* that it also applies to figures with equal (congruent) angles. To this effect, he proves a form of *Elements* VI, 5 (the central part of our *text 10*), implicitly employing Thales' Theorem (*Elements* VI, 2), which itself requires the Parallel Postulate. Leibniz is thus forced to enter the discussion on parallel lines, and here he tries to define these latter as lines which have a common perpendicular. This

---

non pas dans la memoire, car on ne sçauroit retenir les grandeurs, mais dans une mesure materielle gravée sur une regle, ou autre chose. Car si toutes les choses du monde qui nous regardent, estoient diminuées en même proportion, il est manifeste, que pas un ne pourroit remarquer le changement. Par cette definition je demonstre aisement des propositions tres belles et tres generales, par exemple que deux choses estant semblables selon une operation ou consideration, le sont selon toutes les autres; par exemple soyent deux villes inegales en grandeur, mais qui paroissent semblables parfaitement, lorsqu'on les regarde au costé oriental, je dis qu'elles paroistront aussi semblables, quand on les regardera du costé occidental, pourveu que à chaque veue on découvre toute la ville. Cette proposition est aussi importante en Metaphysique et même en Geometrie et en Analyse, que celle du tout plus grand que sa partie. Et neanmoins personne que je sçache l'a enoncée. On demontre par la aisement le theoreme des triangles semblables qui semble si naturel, et qu'Euclide demonstre par tant de circuits" (A II, 1, n. 158, p. 380; A III, 2, n. 79, p. 227–28).

[49] Leibniz thinks that his notion of similarity is wider than the Euclidean one, as it applies also to curvilinear figures (while Euclid's definition in *Elements* VI, def. 1 and *Elements* XI, def. 9 only works for polygons; while he has different definitions for circular sectors in *Elements* III, def. 11, and for cones and cylinders in *Elements* XI, def. 24), and in fact even to non-geometrical entities. In the essay *De analysi situs* (1693), among others, Leibniz attempts to prove that Euclid's definition is just a specification of his own more general notion, and he very much praises the universality of his definition: "Et similitudinum seu formarum consideratio latius patet quam mathesis, et ex Metaphysica repetitur, sed tamen in mathesi quoque multiplicem usum habet ... Causam vero cur similitudinis consideratione non satis usi sunt Geometrae, hanc esse arbitror, quod nullam ejus notionem generalem haberent satis distinctam aut ad mathematicas disquisitiones accommodatam, vitio philosophorum, qui definitionibus vagis et definito obscuritate paribus, in prima praesertim philosophia contenti esse solent ..." (GM V, pp. 179–80). For some theorems that Leibniz wanted to prove with his improved notion of similarity, see below the *note* 67. Others have instead criticized Leibniz' definition as *too metaphysical* and *too general* to be of any use in geometry. Among the classics, see Kant's *Untersuchung über die Deutlichkeit der Grundsätze der natürlichen Theologie und der Moral*, I, § 1 (KGS II, p. 277; but this criticism recurs in Kant), who was referring to Wolff's definition of similarity (dependent, in turn, on Leibniz). In the recent literature, see GIUSTI 1992.

[50] See above, the remarks in *note* 9.

was, as we know, Borelli's definition of parallels;[51] it is good enough, of course, but one should then postulate something like the Parallel Postulate (as Borelli did), or be careful not to implicitly assume that if two straights have *one* common perpendicular, then *every* perpendicular to one of them is also perpendicular to the other (a statement equivalent to the Parallel Postulate).[52] This is what Leibniz does in this text, however, immediately adding that in a quadrilateral with two right angles at the base and two equal sides, the fourth side is equal to the base: another proposition equivalent to the Postulate, and in fact the central theorem that many geometers, from Khayyām to Saccheri, wanted to prove in order to establish it. Finally, Leibniz proves that the interior angle sum of a triangle is equal to two right angles, a straightforward Euclidean conclusion from the (unwarranted) preceding results. In the end, he states that his definition of parallel lines is better than the Euclidean one (the same claim of Borelli), but shows that he intends it as equivalent to that of equidistant lines, and resorts to the real, constructive definition of equidistant lines through motion. This is in fact the first occurrence of the epistemological theme of real definitions in Leibniz' work on parallel lines. The Postulate, however, is really just assumed here rather than any attempt being made to actually prove it.

Around 1683, Leibniz was writing an essay on *mathesis universalis* (*text 11*), or what he calls a mathematical theory of everything falling under the imagination (a *logica imaginationis*).[53] Among many considerations on the subject matter of universal mathematics and its connection with the faculty of imagination (again through the perceptual definition of similarity), a few deal with those mathematical objects that *cannot* be imagined in the proper sense, and can only be symbolically expressed. The primary example is that of imaginary numbers (where "imaginary" here oddly means what cannot be imagined), which can be employed in equations even though they have no geometrical interpretation in real space.[54] But Leibniz wants to add the infinitesimal or infinite quantities that he is using in his new Calculus, as well as the intersection point at infinity of parallel lines. The text does not make any explicit use of the Parallel Postulate, as Leibniz' statements on the matter are all true also in hyperbolic geometry, but the diagram at least betrays a Euclidean conception of the topic.[55]

---

[51] Remember that Leibniz had acquired Borelli's book a few months earlier (in 1678).

[52] Borelli's definition of parallel lines is equivalent to the Euclidean definition only if one accepts the Parallel Postulate (as Borelli did). In hyperbolic geometry, straight lines that have a common perpendicular are in fact parallels in the classical sense (i. e. non-incident); but the reciprocal proposition is false, and there are also other parallel lines (the asymptotic ones) which do not have any common perpendicular.

[53] On the concept of *mathesis universalis* in the Early Modern Age, see CRAPULLI 1969 and RABOUIN 2009; the latter book gives the historical background of Leibniz' project.

[54] Leibniz' example to Huygens is the sum $\sqrt{(1+(\sqrt{-3}))}+\sqrt{(1-(\sqrt{-3}))}=\sqrt{6}$, in which the addition of two complex numbers results in a real number (letter from September 1675; in A III, 1, n. 61, pp. 276–80). Huygens was highly surprised by the example, as the imaginary numbers were hardly employed in the mathematics of the time (A III, 1, n. 62, p. 282). On the topic, see MCCLENON 1923. A recent survey of the literature on the status of imaginary objects and fictional entities in Leibniz' mathematics is SHERRY&KATZ 2012.

[55] It is true, for instance, that the larger the angle ACB, the longer the segments AB; but Leibniz doesn't give the monotonic function of growth, which is different in the Euclidean and hyperbolic

Leibniz also seems to refer again to a theory of infinite but bounded straight lines. In any case, he adds some remarks on the usefulness of intersection points at infinity in the theory of conics, and proposes to call parallelism a "quasi-angle" that cannot be imagined (as it is *not* an angle) but nevertheless may be employed in geometry.

The issues of similarity and imagination are discussed again in a passage of the *Phoranomus* (1689), a long and important dialogue on dynamics that Leibniz wrote during his trip to Italy. Here (*text 16*) he restates the idea that the perceptual definition of similarity is more general than the Euclidean one, as it may be applied to any figure (such as two straight lines or two circles), since it doesn't rest on the notion of an angle. Leibniz touches then cursorily on the notion of parallelism, observing again that the equidistant curve to a conic section is not *similar* to the conic section. The progress made vis-à-vis the earlier statements in this direction (*text 8* above) is that now Leibniz clearly recognizes that (without the Parallel Postulate) the figure equidistant to a given figure may not be similar to that figure (while in 1679 he was only concerned with the fact that the curve equidistant to a parabola is not of the same *genus* of it – a question of algebraic degree). Now Leibniz reconnects the notion of parallelism to that of similarity, which is a big step forward, and of course claims that one should prove that the equidistant line to a straight line is similar to it (as all straight lines are similar to each other).

In the 1680s, finally, we find some marginal reflections on parallelism in Leibniz' commentary on Jungius' system of definitions. Joachim Jungius (1587–1657) was a German logician, mathematician and natural philosopher deeply appreciated by Leibniz, who annotated several of his works and had the opportunity (during a trip to Hamburg in the summer of 1678) to look at the many papers that he had left in manuscript form.[56] Jungius' definition of parallels (*text 15*) clearly shows the marks of the Scholastic logician. He calls simply *inconcurrentes* the non-incident straight lines (the parallels in Euclid's definition), reserving the word *parallelae* for equidistant lines, which are in turn defined through the length of the perpendicular. Since however Jungius also considers straight lines as bounded segments (as it was usual at the time), he remarks that some non-parallel bounded lines may still be *inconcurrentes* (they will only meet if extended). He also remarks that one can use a wider notion of parallelism to mean non-incidence, and this is the case in Euclid. Leibniz however objects that Jungius doesn't give a clear definition of convergence of bounded segments, and seems to propose a notion of direction connected with that of *situs*: so

case. Moreover, he doesn't say explicitly that the two straight lines are parallel only when the angle ACB is a right angle. The diagram, however, shows ACB as a right angle.

[56] During this trip to Hamburg, Leibniz also acquired Fogel's library (and Borelli's book, which was part of it). Martin Fogel (1634–1675) had himself been a pupil of Jungius, and along with his fellow student Johann Vagetius (1633–1691) had published several works of his teacher. Leibniz had a correspondence with both of them, and after Fogel's death Vagetius was able to provide Leibniz with a few other of Jungius' manuscripts. Leibniz' annotations to Jungius are to be found in A vi, 4B, nn. 224–229, pp. 1048–90 (from the years 1678–1680) and A vi, 4B, nn. 241–42, pp. 1211–1306 (from which the *text 15*). Most of Jungius' *Nachlass* was later destroyed by a fire in Vagetius' house. On the relations between Jungius' thought and Leibniz, see Kangro 1969 and the materials in Risse 1977.

that parallel lines are "*paralleliter* situated". The attempt was to be taken up again in the 1690s (see below, *texts 20, 23, 25, 30*). Lastly, Jungius employs *Elements* I, 27–29 to teach how to draw parallel lines in practice (in a field, for instance), connecting his reflections to his own work on *Geometria empirica*, a book published in 1627 that Leibniz had read when he was a student.[57] We should also note that Leibniz kept in his papers a handwritten page by Martin Fogel, in which he reported Jungius' (unpublished) criticisms of Proclus' proof of the Parallel Postulate.[58]

4.4 Leibniz' Studies in the 1690s (*Texts 17–29* and *34*).

In 1689–1690, Leibniz visited Italy and met several mathematicians engaged in researches on the foundations of geometry; among them, he made the acquaintance of Vitale Giordano. This trip had a major influence on Leibniz' work on *analysis situs*, and from the 1690s onward it tends to approach more and more the form of a classical commentary on the *Elements*, indulging less in considerations on the infinite, or a new characteristic, or any other idiosyncratic project on the foundations of mathematics. The first example of this (partially) new attitude is the long *Demonstrationes Euclideas* from 1690, which are the first real commentary of Leibniz on Clavius' *Elements*, and the model for many others in the years to come (culminating with the even longer *In Euclidis* πρῶτα from 1712). The *Demonstrationes* (*text 17*), in fact, do not deal with the Parallel Postulate itself. In Clavius' edition, the proof of the Postulate only appears

[57] Leibniz mentions Jungius' *Geometria* in the *Theodicée*, § 214 (GP VI, pp. 246–47). He probably endorsed many tenets of Jungius' epistemology, and the *Geometria empirica* is one of the many books published at the time that simply had no axioms whatsoever, proving all theorems from definitions alone (Jungius adds, however, a few contructive postulates). Leibniz was also probably influenced by Jungius' definition of a straight line as a self-congruent line, and he reports it as early as 1671 (cf. A VII, 1, n. 1, p. 3); see below *note* 75. Another book by Jungius on elementary geometry that may have interested Leibniz is the *Disputatio philosophica de stoecheosi geometrica* from 1634, which offers strong criticisms to Ramus' geometrical epistemology (criticisms endorsed by Leibniz himself, see above a few quotations in *note* 7 in § 3), but has nothing on the theory of parallels.

[58] Fogel's paper is commenting Proclus' proof as it is found in Commandino's edition of the *Elements*, adding Jungius' remarks. In particular, the observations are dealing with an important Lemma, stating that if a straight line meets another straight line, then it also meets a parallel to it. The Lemma is in fact false in hyperbolic geometry, and may be used as a substitute for the Parallel Postulate. Commandino "proves" it thanks to the definition of parallel lines as equidistant lines. Apparently, Jungius was contending that Commandino had to prove that the distance between the parallel lines is a straight line, asking "si distantia illa numquam est linea recta?". The fragment is too short to give a clear meaning: maybe Jungius just wanted a better definition of equidistance through the constant length of the perpendicular, as he himself was doing in *text 15*; or he may have realized that equidistance itself had to be proved. Then he remarked that Commandino also had to prove that the line that intersects one of the parallels does not "merge" with the second line instead of intersecting it. This may be a hint about asymptotic lines in hyperbolic geometry, and it should be remarked that the impossibility of such a merging was later taken by Saccheri as the main argument to prove the Postulate. Fogel's remarks are however too short to really understand Jungius' mind on the topic. They are to be found in LH XXXV, XIII, 2a, Bl. 117. I thank Siegmund Probst for having drawn this page to my attention and kindly provided a copy of it.

after *Elements* I, 28, since the Jesuit mathematician needed several preliminary results to ground his demonstration; Leibniz' commentary to Clavius, however, is interrupted after the axioms and it doesn't arrive at discussing Euclid's theorems. Some years later, in 1695, Leibniz wrote another long commentary on Clavius (*text 25*), which expands a lot on the system of principles, the epistemology behind the aim of proving all the axioms, the necessity of the identical propositions, and so forth.[59] In this text, he goes so far as to give a proof of all the Euclidean Common Notions, but again fails to address the properly geometrical principles and the Parallel Postulate, abruptly ending the paper with an unfulfilled promise: "… nunc ad principia propriè geometrica ad situm pertinentia veniamus". The essay is still interesting for our topic, however, since Leibniz (before addressing the Common Notions) also discusses the Euclidean definitions, and that of parallels among them. Leibniz says that the definition of Euclid (through non-incidence) is good, but he clearly prefers the one through equidistance; this last, however, "postulates something" because it is not evident that the equidistant line to a straight line should be straight (he has now the counterexample of the conic sections). He briefly discusses the general, analytical notion of an equidistant curve to a given curve, as the curve which is the locus of the centers of the circles touching the given curve. He then remarks that this definition can only be applied in a plane, and fails with spatial curves; thus, we have to add to our definition of parallel lines (as Euclid did) that they are coplanar curves. Finally, so as to get around the difficulties connected with the proof that the equidistant curve to a straight line is itself straight, he proposes another definition of parallels, as lines that have everywhere the same situation the one with the other: this definition was intended to solve all the difficulties connected to the real definition of parallel lines and ground the Parallel Postulate (but

[59] In the text, Leibniz repeats that one can take the Euclidean axioms as unproved hypotheses, or can try to prove them. Here is the transcription of the most relevant passage: "Et revera si quis Euclidea Axiomata neget, ut Franciscum Sanches (autorem libri quod nihil scitur) fecisse accepi, dubitationibus suis ad Clavium missis; fient hypotheses id est hoc saltem habebit Euclides his paucis suppositis concessisque; caetera omnia ab ipso aliisque superstructa fore indubitatae veritates (quodutinam et in philosophia haberemus); sed nunc ipsas illas veritates praeliminares aggrediamur". In these lines, Leibniz mentions Francisco Sánchez' skeptical criticisms to Clavius. These were contained in an undated letter that Sánchez sent to Clavius, that was first published in Iriarte 1940, and may now be read in Clavius' correspondence (Baldini&Napolitani 1992) or Sánchez' *Opera*. In the letter, Sánchez mentions the difficulties in the proof of the Parallel Postulate and states that its demonstration is more obscure that the statement itself (as Arnauld and others would say in subsequent years): "Alia etiam pro indubitatis accipiunt, ut definitions aliquas … et pronunciata, ut decimum tertium primi Euclidis, quod tu cum Gemino et Proclo ab axiomatum numero reiicis; et tamen cum id demonstrare conetur Proclus, petit principum, ut tu putas; et obscurior est et hic et alibi probatio quam id quod probatur". Unfortunately, we don't have Clavius' answer to Sánchez' criticism. Leibniz may have known something of Sánchez objections through his biography, which was published by Delassus in 1636. Leibniz mentions Sánchez' objections several times, always in the context of the possibility of proving the Euclidean axioms. See his own notes to the *Elementa veritatis aeternae*: "Fatendum est tamen nonnulla desiderari posse in Euclidis demonstrationibus, et Franc. Sanchez miserat difficultates quasdam suas ad Clavium, negabatque sibi satisfactum ejus responsis. Sed haec tamen suppleri posse dubium nullum est" (A vi, 4a, n. 113, p. 446; see also the letter to Varignon from February 2nd, 1702, in GM iv, p. 94, where Gerhardt erroneously transcribed Suarez for Sánchez). On the relations among Sánchez, Clavius and Leibniz, see Cunningam&Mellizo 1978 and Olaso 1986.

now we still don't know how, since the proper discussion of the Postulate, as we have remarked, is missing). These remarks show nearly all of what is at stake in Leibniz' writings on the topic in the 1690s: the analytical description of equidistant curves is a recurrent issue in Leibniz' writings of the time (*texts 18*, *19*, *21*, *24*, and later on in *text 34*); a new definition of parallels through the notion of *situs* is implemented in several essays (*texts 20*, *23*, *26*); and the idea that coplanarity is an important feature of the definition of parallels was also to re-emerge very soon (in the long demonstration of *text 31*). We discuss now some of these texts.

In a definition of parallels written in 1690, Leibniz states that the notion of equidistance is more important than that of non-incidence since it can be usefully generalized to curved lines, and he mentions that a family of equidistant curves may be obtained through *evolution* (*text 18*).[60] This notion is then developed in two essays that Leibniz published in the *Acta eruditorum* in 1692 and 1695, in which he explains (*text 19*, in connection to some writings of Johann Bernoulli) that the family of involutes to a given curve (the evolute) is a set of equidistant curves, a theorem proved in fact by Huygens in 1673.[61] Leibniz then adds (*text 24*) that to trace a parallel line (i. e. an equidistant curve) to a given curve one does not need to work at the evolution of the curve, as it is enough to consider the *locus* of the center of a circle rolling on the given curve (as we have seen in the contemporaneous *text 25*),[62] but the construction

[60] I thank Massimo Mugnai, who found this small slip of paper on which Leibniz had annotated this definition of parallels and communicated it to me.

[61] The involute (or evolvent) of a given curve (the evolute) is the curve traced by the free end of an imaginary string which is unwound around the evolute. The iteration the process with strings of different lengths gives rise to a family of involutes, which are equidistant to one another. The inverse transformation, which produces the evolute of a given curve (the involute), may be performed by taking the locus of the centers of curvature of the given curve. The concept may be traced back to the Fifth Book of Apollonius' *Conics* (Propositions 51 and 52), even though Fried&Unguru 2001 rightly points that Apollonius did not *define* new curves through *loci* theorems (as it happens with the modern reading of an evolute). In any case, Apollonius' Fifth Book, which is lost in Greek and only survived in Arabic translation among the Medicean manuscripts guarded by Raimondi (see above, *note* 7 in § 2), was published by Borelli in 1661 (see Giovannozzi 1916 and Guerrini 1999 for any detail about this edition). Following Apollonius' hints, the complete, modern treatment of involution and evolution was given in the Third Part of Christiaan Huygens' *Horologium oscillatorium* from 1673 (pp. 59–90). Here is Huygens' definitions of evolute and involute (even though he doesn't yet have the term "involute"): "Si linea in unam partem cava, filum seu linea flexilis circumscripta intelligatur, & manente una fili extremitate illi affixa, altera extremitas abducatur, ita ut pars ea quae soluta est semper extensa maneat; manifestum est curvam quandam aliam hac fili extremitate describe. Vocetur autem ea, Descripta ex evolutione. Illa vero cui filum circumplicatum erat, dicatur Evoluta". The main result proved by Huygens is that the evolute of a cycloid is itself a cycloid, and he was able to use this theorem to establish that the cycloid is the isochrone curve (so that a pendulum moving on a cycloid takes the same time whatever the arc of oscillation), thus correcting Galileo's wrong guess that the circle was isochrone (*Discorsi*, in *Le opere*, vol. 8, p. 129), and theorizing the first exact clock. Further differential properties of evolutes were later discovered in the eighteenth century, and Monge connected them with projective geometry and geodesic surfaces; these developments link the theory of evolutions with non-Euclidean geometry, but were completely unknown to Leibniz. On Galileo and Leibniz, see Knobloch 1999 and 2012.

[62] The procedure of rolling a circle on a straight line to obtain an equidistant parallel to it was commonly used as a variation on Clavius' construction by the perpendicular motion of a segment. This

with the evolute allows us to draw the parallel curve passing through a given point (as in *Elements* I, 31 for the straight lines).[63] Leibniz, after having generalized the notion of parallels to curved lines, also extends the notion of a rectangle as a (likewise curvilinear) figure with all right angles, and he states that a quadrilateral made by two parallels and their perpendiculars (measuring their distance) is a generalized rectangle.[64] Finally, in an unpublished note from 1693 (*text 21*), Leibniz readdresses the whole question and raises some construction problems in drawing the equidistant through evolution (in case of self-intersecting lines).[65] This last fragment, however, quickly becomes one more reflection on similarity, in which Leibniz repeats that equidistant lines are similar only in regular figures (straight lines and circles, but obviously he needs to prove that), and then he praises the advantages of his new perceptual definition of similarity. It is relevant, in any case, that Leibniz claims that parallel straight lines are not just similar to each other, but also *similarly situated* ("similiter positae"). The concept is important, since it recurs (albeit in different wording) in many papers of the time; but the expression is also very significant, since it resembles Proclus' definition of parallelism as a *similarity in position* (ὁμοιότης θέσεως), a definition that, in fact, was itself aimed to be a proof of the Parallel Postulate.[66]

On the same topic, but no longer in connection with the theory of evolutes, is a Leibnizian proof of Pythagoras' Theorem through similarity that is supposed to improve on Euclid's own demonstration and show the mathematical advantages of establishing similarity as an original (non-derivative) geometrical relation. In the development of the proof, however, Leibniz does not discuss the Parallel Postulate, just assuming it as an "evident truth" (*text 28*).[67]

---

kind of construction was also given by Saccheri (*Euclides vindicatus*, Scholium 1 after Proposition 37; pp. 92–94) to prove the Parallel Postulate, even though he blundered here in a bad limiting process connected to the celebrated problem of Aristotle's Wheel (see Drabkin 1950). But even Leibniz, who used infinitesimals much better than Saccheri, was confused by similar arguments (see below, *text 34*, also connected with the issue of evolution).

[63] Given a curve, one can construct its evolute and then recover the curve and all its parallels as involutes of the evolute.

[64] This is a true result for equidistant lines, but not for parallels as non-incident straights, as is clear from Saccheri quadrilaterals.

[65] The general problem is that a curve parallel to a given curve has a cusp when the radius of the circle describing the parallel is equal to the radius of curvature of the given curve. This happens when the equidistant curve touches the evolute.

[66] The passage is Proclus, *In Euclidis* 373. Proclus claims that the property of transitivity may only be attributed to a restricted class of relations: and similarity among them. Since parallelism is a kind of similarity, then, it has to be transitive (i. e. *Elements* I, 30 has to be true). As *Elements* I, 30 depends on the Parallel Postulate, this amount to a proof of the Postulate itself. Note that the expression "similiter positae" may also have reached to Leibniz from other sources, and in particular Archimedes' Fifth Postulate in *De aequilibris planorum*, which uses the expression ὁμοίως κεῖσθαι.

[67] Note that in *text 21*, Leibniz' example of a theorem that can be easily proved by similarity even though it refers to measure theory, is that of *Elements* XII, 2, stating that circles are to one another as the squares on the diameters. This proposition is quite important in a foundational context, as it is the first one in which Euclid employs the method of exhaustion. Leibniz' stance that it is useless to resort to infinitary procedures to prove it, and it is better to envisage a new definition of similarity (his own) that could be applied to curvilinear figures, is often restated. See for instance the

We should also mention, among the developments of Leibniz' and Bernoulli's use of the evolution in mathematics, the attempt at the solution of the brachistochrone problem given by Joseph Sauveur[68] in 1697; this attempt attracted Bernoulli's criticisms, who complained with Leibniz that Sauveur had not understood the new infinitesimal methods.[69] Leibniz largely agreed with Bernoulli's opinion, even if for different reasons, and in the course of their exchange on the question (which discloses interestingly different foundational perspectives on infinitesimals), Leibniz blamed Sauveur for having considered two lines making an infinitesimal angle as parallels in the proper sense (*text 27*).[70] It is remarkable that in an age in which infinitesimals

---

Leibnizian proofs of *Elements* XII, 2 in *De rebus in scientia mathematica tractandis* (1680–1682; in A VI, 4A, n. 92 pp. 380–81), *Circa geometrica generalia* (1682; in MUGNAI 1992, pp. 145–46), *De analysi situs* (1693; GM V, p. 182), *Specimen geometriae luciferae* (1695; GM VII, pp. 276–77), *Scientia mathematica generalis* (1700; in DE RISI 2007, pp. 625–26), *Initia rerum mathematicarum metaphysica* (1715; GM VII, p. 24). As for Pythagoras' theorem, demonstrations of it through similarity were probably known already in Antiquity, and some scholars have suggested that they were in fact known to Euclid (see MUELLER 1981, pp. 172–73). Leibniz gives several proofs of it, one being in a letter to Schenk from 1712 touching several issues on the foundations of geometry (partially published in DE RISI 2007, p. 621).

[68] Joseph Sauveur (1653–1716) was a French mathematician especially praised, at the time, for his teaching skills. He composed a treatise on elementary geometry that enjoyed wide manuscript circulation among professors of mathematics, and was only published by Le Blonde in 1753. Sauveur's *Geometrie* follows in many respects Arnauld's *Nouveaux elemens*, and has a section on parallel lines that is even simpler than its model, since it takes parallel lines as equidistant straights and uncritically proves several theorems about them (including the Parallel Postulate); other theorems are not even proved, as they should be self-evident enough (pp. 74–79). We have no hint whether Leibniz may have known this book; the present discussion, in any case, is not concerned with it.

[69] The brachistochrone problem was raised by Johann Bernoulli in 1696 and amounted to finding the trajectory of fastest descent of a heavy body between two given points. Bernoulli showed (from Huygens' theory of evolution) that the cycloid is also the brachistochrone curve. Leibniz discussed the issue as well, acknowledging Bernoulli's correct solution (see his *Solutionum problematis curvae celerrimi descensus*, in GM V, pp. 331–36).

[70] Sauveur's solution to the brachistochrone problem was given to Bernoulli by L'Hospital, who was unable to detect any mistake in it, and Bernoulli sent it to Leibniz, along with his remarks, on January 29th, 1697 (A III, 7, nn. 68–69, pp. 265–75). In this letter, Bernoulli explicitly notes Sauveur's treatment of parallels: "… et deinde peccat in principia calculi differerentialis, quando considerat duas lineas angulum infinite parvum constituentes ut absolute parallelas" (p. 268). Leibniz replied with a long letter on February 8th, defending Sauveur from some of Bernoulli's criticisms but still finding in his attempt an *elegans & subtilis aberratio*, an instance of which is that on parallelism. In particular, since the center of the circle of curvature of a curve at a point is defined by the intersection of two infinitely close normals to the curve, considering these as parallels would destroy the very possibility of measuring the curvature. Leibniz adds, however, that this very theoretical mistake does not hinder Sauveur's proof as Bernoulli claimed, and the real error is to be found elsewhere: "Quod attentatam a Domino Salvatore solutionem attinet, equidem concedi potest, non tantum mediam geometricam et mediam arithmeticam duarum quantitatum infinitesimaliter, seu per inaccomparabile, differentium coincidere, sed et duas reetas angulum infinite parvum facientes haberi posse pro parallelis, cum de alia recta eas secante quaeritur, et (quantum judicare possum) Dominum Salvatorem his regulis male usum non esse. Sed alia sunt, quae solutioni ejus obstant …". The exchange continued further, with a long answer by Bernoulli again complaining about Sauveur (from March 2nd; in A III, 7, n. 74, pp. 305–12), a reply by Leibniz closing the discussion (March 5th; in A III, 7, n. 75, pp. 312–15), followed by some apologies by L'Hospital for not having

were sometimes equated with zero (and sometimes not), Leibniz insists on the point; remember however that already in the 1680s Leibniz considered parallels as forming an imaginary quasi-angle (*text 11* above). Some years later, in a paper written to philosophically defend the new Calculus, Leibniz would praise the fruitfulness of the continuity principle as an heuristic tool in the art of discovery, saying (among other things) that parallels could be considered, *by continuity*, a special case of incidence (*text 29*). The reference seems clearly to be to his previous studies on projective treatments of conic sections.[71]

The last item of this collection of studies is the essay *De lineae super linea incessu*, which Leibniz published in the *Acta eruditorum* in 1706 (*text 34*). It was too quickly written and reveals an obvious mathematical slip. Leibniz attempts a general theory of the movement of a curve along another curve, in which the two curves always remain tangent. He attempts to split this general motion into a translational component of the rigidly moving curve (that he calls a *radent* motion) and a rotational one of the same curve (the *provolution*); the final aim would be to provide a theory that can encompass the evolutions and involutions as special cases of curves produced by this kind of composite motion. Leibniz' description of radent motion, however, was grossly defective, as was (privately) remarked by Johann Bernoulli and Jakob Hermann; Leibniz acknowledged the mistake.[72] Our interest in the affair is that, in describing the flows of points in radent motion, Leibniz imagines (sim-

seen Sauveur's mistake (to Leibniz, March 17th; in A III, 7, n. 81, pp. 331–33) and Leibniz' polite answer to these (March 25th; in A III, 7, n. 83, pp. 340–44). But still one year later, in commenting on a mathematical error by James Gregory, Leibniz said to Bernoulli that Gregory had run into the same thorns as Sauveur before him (September 30th, 1698; in A III, 7, n. 233, p. 908). A technical reconstruction of the debate on Sauveur's solution to the brachistochrone is in KNOBLOCH 2002 [C].

[71] In *text 29*, Leibniz mentions his own *Lettre sur un principe général* which appeared in the *Nouvelles de la Republique des lettres* in July 1687. In it, he formulated a quasi-mathematical definition of continuous functions that is not too distant from the modern Weierstrassian approach through ε and δ. Here is the main point: "On le peut enoncer ainsi: Lorsque la difference de deux cas peut estre diminuée au dessous de toute grandeur donnée in datis ou dans ce qui est posé, il faut qu'elle se puisse trouver aussi diminuée au dessous de toute grandeur donnée in quaesitis ou dans ce qui resulte" (GP III, p. 52; cf. also the *Principium quoddam generale*, in GM VI, p. 129). Note that the last sentence on the vanishing moment of an infinitesimal quantity recalls Newton's famous definition of it in the scholium to Lemma 11 in Section I of Book One of the *Principia*, which says that the ultimate ratio of vanishing quantities is to be understood neither before they vanish nor after they have vanished, but at the very instant when they vanish (NEWTON, *Principia*, vol. 1, p. 87). Note also that in the same *Lettre*, Leibniz employs the principle of continuity to justify the similar mathematical treatment of the different conic sections, even though in this case he doesn't mention the parallels converging at infinity: "L'on sçait que le cas ou la supposition d'une Ellipse se peut approcher du cas d'une parabole autant qu'on veut, tellement que la difference de l'Ellipse et de la parabole peut devenir moindre qu'aucune difference donnée, pourveu que l'un des foyers de l'Ellipse soit assés eloigné de l'autre, car alors les rayons venans de ce foyer eloigné differeront des rayons paralleles aussi peu que l'on voudra, et par consequent tous les theoremes Geometriques qui se verifient de l'Ellipse en general, pourront estre appliquées à la parabole, en considerant cellecy comme une Ellipse dont un des foyers est infiniment eloigné ou (pour eviter cette expression) comme une figure qui differe de quelque Ellipse moins qu'aucune difference donnée" (GP III, p. 52).

[72] Letters by Johann Bernoulli from May 13th and 22nd, 1706 (GM III, pp. 785–87 and 790–93). Leibniz' answer is (partially) reproduced as *text 34*.

plifying a bit) drawing a curve by the parallel motion of a segment along another curve. The flow of the free end of the segment traces a curve which is not equidistant to the first (it may even intersect the original curve), but is congruent with it. It is, in fact, just a translation of the original curve. The justification that Leibniz gives to the congruency of the curves is (roughly speaking) that the segment may be very short and even infinitesimal, so that the original curve and the one produced by the flux of the point cannot but be congruent. One sees that Leibniz is just one step away from saying that the equidistant line to a straight line (which is generated by such a motion) is congruent with it and thus straight itself. He refrains from making this statement, though, and does not add mistake to mistake; he immediately quotes, however, Hobbes' generalized theory of parallels. The text has some relevance for our problem, in any case, since it shows the final outcomes of Leibniz' reflections on the evolute, as well as how blunders on infinitesimals were easy to commit at the time.

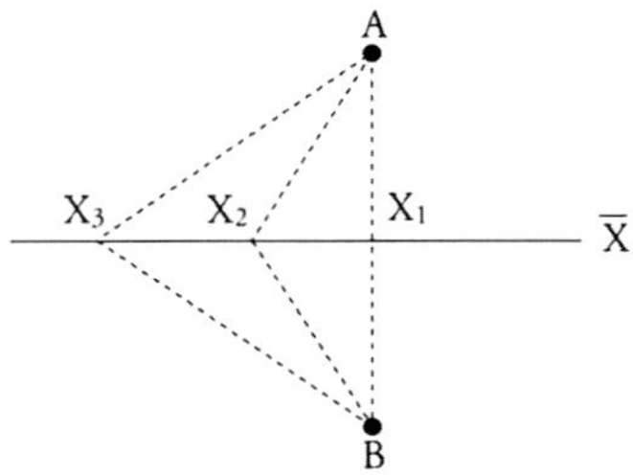

The other important trend in Leibniz' studies on parallel lines in the 1690s is their new definition in relation to situation, and its symbolic expression. We have seen the sketch of such a definition in an essay on Clavius (*text 25*), but it also appears in the *Elementa geometriae generalia* (*text 23*); it says that two parallel lines are related to one another everywhere in the same way ("*ubique eodem modo se invicem habent*"), that is, they have everywhere the same reciprocal situation.[73] This definition of parallels is based on the notion of situation and congruence, and mirrors the analogous definition of a straight line as a "*sectio plani utrinque se habens eodem modo*", i. e. a section of the plane which has the same situation on both of its sides. This last definition of a straight line is expressed by a few equivalent formulas in Leibniz' new formalism for geometry. One of them is that $\bar{X}$ is a straight line if $A.X \simeq B.X$. In the symbolism of the *characteristica geometrica*, letters denote points, letters with a superscripted line the set of all points satisfying a given condition, a dot means the reciprocal situation between two figures (which is simply the distance between them), and the symbol $\simeq$ denotes congruence. Thus, the formula above means that if points X (say X_1, X_2, X_3, …) have the same situation (distance) to two given points A and B, then the set of all points X, that is $\bar{X}$, is a straight line.[74]

---

[73] The implicit reference to situation in the above expression is made explicit in the analogous definition of a straight line (that we are going to see immediately). The way (*modus*) in which two geometrical figures are related the one with the other (*se habent*) is *situs*. Situation, in fact, is often defined by Leibniz as the *modus* of space (see *Initia rerum mathematicarum metaphysica*, GM VII, p. 18). The expression may also be taken as synonymous with the "*similiter positae*" that we have seen in *text 21*.

[74] This formula only works in a plane, of course, and to define a straight line in space one has to fix another point C to which all the X's have the same situation. In space, however, the above formula defines a plane, and Leibniz employs it to this effect. The same definition of plane recurs in

Another formula that Leibniz employs to the same effect is that if $\bar{X}$ and $\bar{Y}$ are segments of a plane (half-planes, in fact), and the set $\bar{Z}$ is their common section (which means that every Z is a certain X and every Z is also a certain Y) then the section $\bar{Z}$ is a straight line if $\bar{Z}.\bar{X} \simeq \bar{Z}.\bar{Y}$. These formulas should express the property of a straight line of being identically situated with respect to its two sides, and should be connected to Leibniz' definition of a straight line as an axis of rotation (as they can be easily extended to the three dimensions).[75]

Similarly, Leibniz' definition of parallel lines can be formalized stating that if $\bar{Y}$ and $\bar{X}$ are straight lines, and A is a point on $\bar{Y}$ (that is, a certain $Y \in \bar{Y}$), then $\bar{Y}$ and $\bar{X}$ are parallel if the situation of A with the straight line $\bar{X}$ is the same as the situation of any Y with $\bar{X}$, that is, if $A.\bar{X} \simeq Y.\bar{X}$.

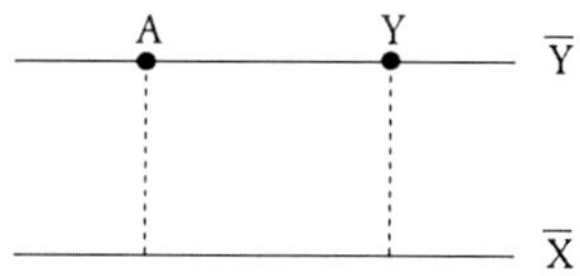

This formalism is firstly used as an explanation of the definition of parallel lines in a fragment from 1692 (*text 20*), but it becomes recurrent in the following years (*texts 26, 38*). It should be noted that since the situational relation (expressed with the dot) is always taken by Leibniz as a distance, this symbolic definition simply amounts to that of the equidistance of the two lines. It should also be noted that since the only notions employed in the aforementioned symbolic definition are those of point, situation, and congruence, which are *simple* (i. e. not complex) *ideas* in Leibniz' logical construction, this formula may be considered an *essential definition* of parallel lines. Leibniz adopted this formal definition in order to accommodate his general notion of parallelism as equidistance (a notion that that can be extended to curvilinear figures) to the new geometrical calculus – as it is explicitly stated in later essays (cf. *text 38*, where the set of points parallel to another set may be a circumference or a plane or a sphere).[76] Once again, Leibniz should prove that if the set of points $\bar{X}$ is a straight line, then the equidistant set $\bar{Y}$ is also a straight line. At the end of the first essay on the topic (*text 20*), he offers in fact some scanty considerations on the motion of the perpendicular segment producing it (as in Clavius). Already there, however, he also explores a quasi-perceptual interpretation of the

---

Lobachevsky (*New principles*, ch. 2, § 18; see also above, *note* 8); and a similar one in Gauss (see again Zormbala 1996 for a complete discussion).

[75] This is a very special case of Leibniz' formalism for geometry, which is quite complex and underwent several changes over the years. For a more complete presentation, see the references in *note* 1 in § 1. The first formula for the straight line can be found almost everywhere in Leibniz' writings on the subject, and in particular in the *Analysis geometrica propria* (GM v, pp. 172–78) from 1698, as an explanation of the above-mentioned definition of the "*sectio plani utrinque se habens eodem modo*" (GM v, pp. 174). The second one comes from the *In Euclidis* πρῶτα, GM v, p. 186. About this last formalization, we should note that in some papers Leibniz would like to define a straight line as a line that bisects the plane into two congruent parts; to this definition, already Giordano objected that many curves divide the plane in congruent parts (think to $y=x^3$). A possible Leibnizian source is Jungius' *Geometria empirica*: "Recta linea dr. linea cujus partes versus eandem plagam sibi mutuo applicatae invicem congruae sunt, sive dr. linea cuius una plaga non differt ab altera".

[76] Note that in *text 20* the section on symbolic expressions does not refer to straight lines, and seems to offer a general formalism for equidistant lines. Leibniz says, however, that the intended interpretation of it is about straight lines.

ubique eodem modo se invicem habere. Leibniz imagines that several observers are situated at different points on $\bar{X}$, and maintains that they will all perceive $\bar{Y}$ in the same way. This means that the set of points Y have a uniformity property of some kind (with respect to the straight line $\bar{X}$), by which it appears situationally identical (or rather, indiscernible) from all the points of views (or perspectives) located on $\bar{X}$. This kind of uniformity gives to the set $\bar{Y}$ the unitary structure of being *a parallel line to* $\bar{X}$. The same is true the other way round (taking $\bar{Y}$ as a collection of points of views, and $\bar{X}$ as a perceived object).

In an important essay from 1695 (*text 26*), Leibniz was to try to exploit this phenomenological characterization of parallel lines, in connection with the combinatorial techniques of his *characteristica geometrica*, to prove the Parallel Postulate. He starts with his new definition of parallels ("*ubique eodem modo se habere*"), without characterizing them through equidistance or the above-mentioned formal definition (which amounts again to equidistance), as in his previous studies. On the contrary, Leibniz immediately characterizes the definition perceptually and imagines that an observer moves along the straight lines $\bar{X}$ looking at the line $\bar{Y}$, while another observer moves along $\bar{Y}$ looking at $\bar{X}$, both of them being in uniform motion. He remarks that everything is the same: the two observers perceive no change while moving (each of them always look at the same straight line from equivalent points of view), nor are the perceptions of one observer different from those of the other. This should be the very meaning of the definition of parallels. The grounds of this indiscernibility may be found in the similarity of all straight lines to each other; but also in their reciprocal position (situation), that we have to investigate. Thus, if we consider metric properties and distances,[77] we should say that in the case where $\bar{X}$ and $\bar{Y}$ are parallels (under this definition), and if the first observer moves from X_1 to X_2 with uniform motion, and the second from Y_1 to Y_2 again uniformly (and thus the lengths of their trajectories are equal, $X_1X_2=Y_1Y_2$), then we have: $X_1.Y_1.\bar{X}.\bar{Y} \simeq X_2.Y_2.\bar{X}.\bar{Y}$. This last formula should express the peculiar property of symmetry that we find in parallel lines, stating that the reciprocal situations of the two observers to one another and in relations to the two whole lines that they are tracing (their trajectories) are indiscernibles, and thus congruent ($\simeq$). Starting with this key formula arrived at by his consideration about indiscernibles, Leibniz goes on to state a common axiom of his *characteristica geometrica*, that similar determinants produce similar determinates.[78]

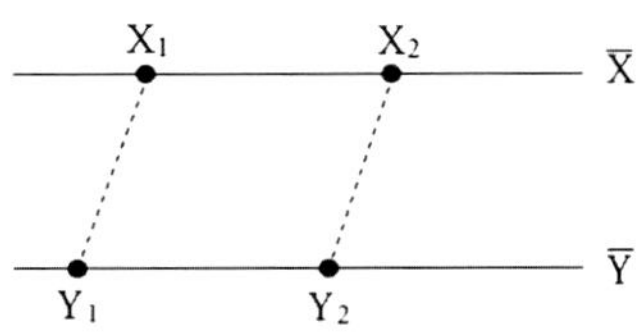

[77] Leibniz seems to think that if we stay with the consideration of similarity among straight lines, we don't need to consider their reciprocal position: all straight lines *se habent eodem modo*, simply *qua* straights. But if we move to metrical geometry in the proper sense, then their reciprocal position is also relevant, and thus two straight lines *se habent eodem modo* in every respect only if they enjoy a peculiar situation the one with the other: if they are parallel.

[78] The principle is widely employed, for instance, in the *Initia rerum mathematicarum metaphysica* (GM VII, pp. 17–29) and it is one of the bases of Leibniz' combinatorial rules for his new symbolism. We may mention that Leibniz sometimes uses this principle to sketch a few demonstrations of basic Euclidean theorems; for instance, he states that the superposition employed in *Elements* I, 4

Then he goes on, claiming that the above formula can be reduced to $X_1.Y_1 \simeq X_2.Y_2$, through a kind of simplification of equal situational relations (eliminating $\bar{X}.\bar{Y}$ from both sides, thanks to the axiom), and the congruence itself implies the equality of the lengths of the segments: $X_1Y_1 = X_2Y_2$; which means that the lines are equidistant. Thus, starting from the definition of parallel straight lines as lines that have everywhere the same reciprocal situation, passing through some phenomenological considerations about the indiscernibles and a couple of combinatorial passages, Leibniz proves that two such lines are in fact equidistant. This amounts (as we know) to a kind of proof of the Parallel Postulate, applied to straight lines *identically situated* (rather than of to non-intersecting lines). In fact, Leibniz immediately deduces (again through combinatorics) the parallelism of the two equal transversals X_1Y_1 and X_2Y_2 and the identity of the internal and external angles of these with X_1X_2 (*Elements* I, 29, equivalent to the Parallel Postulate), the existence of rectangles (a statement again equivalent to the Postulate) and Playfair's Axiom on the uniqueness of the parallel line through a point. He concludes with a restatement of the formal definition of parallel lines (as in *text 20*, but now in a way that entails the Parallel Postulate), saying that if $\bar{X}$ is a straight line and A a point outside it, then if $A.\bar{X} \simeq Y.\bar{X}$, the set $\bar{Y}$ is the straight parallel line passing through A.

If we accept the whole line of reasoning, Leibniz has proved that two straight lines which are parallel by his new definition are also equidistant; and thus the Parallel Postulate can be deduced from their definition (as it should be, Leibniz thinks). Once again, Leibniz should prove (as in the case of the definition through equidistance) that straight lines enjoying this new property of parallelism are possible and real. It should be clear, however, that Leibniz' final goal is now closer, as he has a definition of parallel lines which is half-mathematical (or kinematical) and half-perceptual, and it fits very well with his plans of proving the Postulate through a property of space itself. In other words, the internal possibility of straight lines having those (still rather vague) properties of uniformity needed to ground the Parallel Postulate seems to follow from a property of symmetry of space itself, or perhaps (which is not too different for Leibniz) from a feature of our perceptual abilities. It is interesting to remark, even though Leibniz does not further develop this connection with physics and phenomenology, that the symmetry of space hinted at by his definition of parallel lines is that two observers in inertial motion (e. g. uniform and straight motion) may have such trajectories (e. g. parallel lines) as to be reciprocally undistinguishable.[79]

Finally, we have to briefly discuss Leibniz' *Remarques sur les nouveaux elémens de géometrie* (*text 22*). This is a manuscript from 1693 (or around then) in which

should not bother us, since "quae aequalibus eodem modo determinentur, aequalia sunt", and the two superimposed triangles are determined by equal features; similarly, discussing *Elements* I, 5, which proves that the angles B and C of an isosceles triangles are equal, Leibniz remarks: "cum enim in constructione figurae B et C se habeant eodem modo, etiam quae inde resultant quoad B et quoad C se eodem modo habebunt" (both quotes come from Leibniz' marginal notes to Clavius' *Euclidis*).

[79] Leibniz' above-mentioned proof is that such indiscernibility only amounts to the equidistance of the trajectories, and thus to the fact that inertial paths are Euclidean. This last reduction of indiscernibility to equidistance, in the end, is grounded on Leibniz' very concept of space as a simple system of distances.

Leibniz comments on Arnauld's work on geometry. Leibniz' remarks concentrate on Book Five of the *Nouveaux elémens*, which deals with the principles of the First Book of Euclid's *Elements* and simply states the Parallel Postulate as an axiom; Arnauld tried to prove it in Book Six, but Leibniz left no remarks on this. In commenting on the Parallel Postulate as it appears in Book Five, however, Leibniz attempted his own demonstration of it: a demonstration which bears no connection with Arnauld's proof in Book Six, but accepts Arnauld's formulation of the Postulate as stating that two approaching lines will meet.[80] Leibniz wrote two different drafts of his reading notes to Arnauld, and his treatment of the Parallel Postulate differs greatly in the two versions. The first draft seems to rely on a principle of continuity grounded on the self-similarity of the straight lines. Since straight lines are self-similar and similar to one another, if their distance continuously decreases, they will finally meet. The rationale of Leibniz' proof is not completely clear, and it simply puts aside the possibility of asymptotic lines, basing this claim on a property of self-similarity. Strictly speaking this may be true, since similarity is only well-defined in a Euclidean space; and yet, one would like to read a more detailed proof. In any case, this Leibnizian attempt did not survive for long, and Leibniz' second draft of his notes proposes a completely different demonstration. The second proof is somehow eccentric with respect to the main lines of investigation on the Parallel Postulate in the 1690s, and closer to his earlier reflections on infinity. It begins by stating that a straight line cuts the plane into two congruent parts,[81] and then goes on to say that if two lines were parallels, then a half-plane individuated by the first line would contain a half-plane individuated by the second line plus the infinite strip cut by the two parallel lines. Leibniz then subtracts the infinite strip from a half-plane and adds it to the other one to find a contradiction with the hypothesis that both lines were cutting the plane into congruent parts. Thus, the general assumption of parallelism is false, and the lines should meet. The main point is again that Leibniz doesn't want to give up the Common Notion that the whole is bigger than the part even in the infinite case: and now (as in the 1670s) this claim leads him to paradoxical results.[82] In the 1670s (*texts 3* and *9*), however, he intended to escape these consequences by denying that every straight line cuts the plane into equal parts, and trying to maintain that there is only one line (among parallels) which divides the plane into two congruent half-planes (the "central line"). But now, he would like to endorse his own (new) definition of a straight line as a line which behaves identically with respect to its two sides, a *sectio plani utrinque se habens eodem modo*. In other words, he thinks that the symmetry property that characterizes a straight line in space has to entail the indiscernibility (and thus the congruence) of the two half-planes individuated by it or by every other

[80] See above Arnauld's exact formulation at *note* 3 in § 3. On Arnauld's work as a geometer, see Bopp 1902; on his relations with Leibniz, Bopp 1929. The latter text provides a transcription of Leibniz' notes to Arnauld, which however skips (without warning the reader) Leibniz' proof of the Parallel Postulate.

[81] See above, *note* 75.

[82] A clear statement to this effect is in the already quoted letter to Gallois from 1672; in A III, 1, n. 2, p. 11.

straight line (as all straight lines are similar and congruent to each other). Moreover, Leibniz has already devised (*text 20*, from 1692) his perceptual definition of parallels as perfectly indiscernible lines. Most of all, since the 1670s Leibniz has developed a full-fledged conception of space as a system of relations. In sum, in 1693 Leibniz has to give up the idea held in his youth of straight lines differently situated in absolute space which do not cut congruent half-planes. He has to conclude the present reasoning, thus, by just stating that parallel lines are a contradiction. He probably realized that he had proved *too much*, as this argument would preclude the existence of parallels altogether, and not just prove that *converging* lines (a hypothesis never used by Leibniz in the proof) will meet. He could possibly have played with the idea of parallel lines meeting at infinity, but he never develops the topic any further, and the draft of this proof seems just a clumsy dead end in Leibniz' researches on the Parallel Postulate.

4.5 Leibniz' Late Studies (*Texts 30–39*).

After the turn of the century, Leibniz' writings still show some discussions on parallels and several attempts to ground Euclid's *Elements*. We have already mentioned a place in the *Nouveaux Essais* in which Leibniz discusses real definitions through the example of parallel lines (*text 33*). The text is relevant for Leibniz' epistemology, but adds nothing new from the strictly mathematical point of view, only restating the definition through equidistance (and motion). A similar definition is also to be found in a contemporary collection of definitions (*text 32*). In the same years, Leibniz continues to write commentaries on Clavius, and one of them (*text 30*) restates the general program of proving all the axioms, and that Clavius' particular demonstration of the Parallel Postulate is defective as it assumes that the equidistant to a straight line is straight. Leibniz then sketches a new proof, but it is too short to be properly understandable and in any case it seems to prove *Elements* I, 28 (a statement independent from the Postulate) through some property of symmetry. The only relevant feature of the proof is that Leibniz' line of reasoning seems to be that two straight lines that make equal angles with a secant share the same situation, and thus are parallel. This is one of the few texts in which Leibniz points to the idea of some kind of direction theory of parallel lines, saying that parallel lines have congruent situations.[83] Another

[83] The very concept of "congruent situations" (*situs unius alteri congruus*) is a bit vague, and it seems that any two congruent figures (and any two equal straight lines) should have congruent situations. In any case, while in other texts Leibniz claims that parallel lines have the same *reciprocal* situation (a complex notion that involves the consideration of each point of each line, as we have seen), here he seems to point to an absolute concept of situation that lines may share; the latter might be characterized as their common direction. A direction theory of parallelism, stating that parallel lines are lines in the same direction, was to be discussed later in the eighteenth and nineteenth century. In this context, Leibniz' *analysis situs* was often to be quoted, saying that two lines sharing the same direction have the same *situs*. Our *text 30* is one fragment that may somehow approach (but not much) this notion of parallelism, which is otherwise quite absent from Leibniz' thought. We may notice that the notion of direction played a significant role in the development of spatial notions in the early seventeenth century, when a "centered" world view was superseded by

discussion of Clavius dates from 1712 (*text 35*), where Leibniz quotes Proclus' proof and shows that he had read Nasīr ad-Dīn's ("too lengthy") demonstration in Clavius' commentary.[84] The important essay *Spatium absolutum* from 1714 (*text 37*) states again the definition of parallels as straight lines that have the same reciprocal situation (the main definition of the 1690s), immediately concluding that they are equidistant and thus that different pairs of parallels have the same situation with one another (i.e. they form Euclidean quadrilaterals). An almost contemporary study (*text 38*) reformulates the definition in symbolic language, extending it to parallel circles, planes, and spherical surfaces. Leibniz' definition, in fact, applies to equidistant lines: in Leibniz' formalism, the line $\bar{Y}$, which is the locus of all points Y that have the same situational relation with $\bar{Z}$ (and thus satisfy the formula $Y.\bar{Z} \simeq B.\bar{Z}$), is simply the curve equidistant to $\bar{Z}$ and passing through Y and B. When he further discusses (in the following paragraph of the same text) the locus which is *uniquely determined* by Y (now called A) and B, on the other hand, he is pointing to the straight line passing through A and B, which might be not an equidistant line (unless one proves the Parallel Postulate). The following combinatorial manipulation and its conclusion (the *consequentia memorabilis*), that attempts to prove the equidistance of this line and the symmetry of parallelism, is clearly mistaken. Lastly, another paper on the foundations of geometry (*text 39*) states the classical theorem that one circle passes through three non-collinear points.[85] The construction by Euclid implicitly assumes

---

a general conception of space as an "affine" structure organized through parallel, rather than radial, directions; on the topic, see the remarkable Miller 2014.

[84] We may remark that Leibniz follows Clavius in raising some doubts about Proclus' proof, since it assumes the so-called Principle of Aristotle, stating that two incident straight lines will diverge so as to become more distant than any given quantity (Leibniz underlined the relevant passage in Clavius' commentary). This "axiom" was in fact quoted and employed by Proclus, who accepted it on authority, interpreting quite liberally an Aristotelian passage in *De Caelo* A 5, 271b30-32. Commandino accepted Aristotle's principle as a proper axiom (*Euclidis*, p. 19v). Clavius thought that this principle has to be proved, otherwise one would not prove the Parallel Postulate, but just substitute it with the Aristotelian principle. He attempted to prove it, in fact, but was only able to show that two crossing straight lines will diverge more and more; not, that their distance will become larger than any given distance. For this reason, Clavius accepted in the end the proof of the Parallel Postulate given by Nasīr ad-Dīn (which does not rely on Aristotle's principle), instead of the one by Proclus. Leibniz may have been perplexed for the same reason. Aristotle's principle, in fact, is not crucial for the demonstration of the Parallel Postulate, and it was *proved* in all its strength by Saccheri (*Euclides vindicatus*, Proposition 21, pp. 28–29), who however realized that it is not strong enough to substitute the Postulate (Proclus had reached his "demonstration" of the Postulate thanks to a paralogism, not thanks to Aristotle). The principle of Aristotle, in any case, has several interesting relations with the Parallel Postulate in weaker axiomatizations: see Greenberg 1988 and Pambuccian 2006.

[85] A similar statement is already to be found in Leibniz' *Introductio ad scientiam generalem* from 1679: "Exempli gratia quia non nisi unicus circulus per tria puncta *A, B, C* duci potest, sequitur illis tribus punctis datis quaesiti circuli, determinate posse reperiri quaesiti circuli centrum, seu punctum quod eodem modo se habeat ad puncta *A* et *B* et *C*, quod fiet, si tam ex medio ipsius *AB* educatur perpendicularis *FG*, hujus enim quodlibet punctum aeque respiciet tam punctum *A* quam punctum *B*, et ex medio ipsius *BC* educatur perpendicularis *HK*, cujus iterum quodlibet punctum eodem modo respiciet puncta *B* et *C*; ergo si duae perpendiculares se secent in *D* (quod fit si non sint parallelae seu si *A, B, C* non sint in eadem recta), punctum commune (in quo solo se secare

the Parallel Postulate stating that the bisectors of two segments drawn between the three given points actually meet in one point (which is the center of the circle); an assumption that was already made explicit by Vitale Giordano and others.[86] Leibniz remarks that in the case of three collinear points the bisectors would be parallel, their point of intersection (the center of the circle) lying at infinity. In this fragment, the Parallel Postulate is just accepted, but the issue has nonetheless some relevance as some later mathematicians (such as Farkas Bolyai) wanted to assume the existence of a circle through three non-collinear points as a viable substitute for the Parallel Postulate.[87]

The longer essay on *Calculus situs* is another commentary on Clavius and addresses again the notion of parallels, discussing the relation between Euclid's definition as non-incident lines and Leibniz' definition of equidistant lines (*text 31*). The main point of the discussion is that Euclid's definition of parallels characterizes them as lying on the same plane. This had been recognized since Antiquity as a necessary condition for parallelism, since Euclid employs slanting non-intersecting lines in the stereometric books of the *Elements*. On the other hand, it seemed that equidistant straight lines are obviously in the same plane, and thus someone else (relying on a different definition of parallelism) claimed that Euclid's own definition was redundant.[88] Leibniz finds the definition defective, as it assumes coplanarity, while Euclid ought to have proved its possibility (to get a real definition). In fact, Euclid proves the coplanarity of intersecting lines but has no similar treatment for parallels.[89] The problem is however even more relevant when applied

possunt) aeque respiciet puncta *A* et *B* et *C* et proinde erit centrum" (A vi, 6a, n. 89, p. 371). The theorem is also formalized in Leibniz' *characteristica geometrica*, as many of his formulas deal with the combinatorial definition of a circle, and it is always determined by three points. Leibniz also worked on a general theory about the number of points needed to completely determine an algebraic curve of any given degree, the solution of which is normally ascribed to James Stirling in 1717 (the year following Leibniz' death), even though the main results were already known in the seventeenth century.

[86] The Euclidean theorem is *Elements* IV, 5, which teaches how to circumscribe a circle to a given triangle. Euclid proves that the point of intersection of the bisectors is the center of the circle, but not that this point exists, even though this is an immediate consequence of the Parallel Postulate. Clavius noted that something was missing, but he didn't quote the Parallel Postulate (*Euclidis*, p. 153), while Giordano applies it explicitly in the construction (*Euclide*, p. 153).

[87] See F. Bolyai, *Kurzer Grundriss*, p. 45. The statement is however recurrent in the researches on the Parallel Postulate, and it is not Bolyai's invention. In hyperbolic geometry, given three non-collinear points three mutually exclusive possibilities are given: they lie on a circle; or they lie on a hypercycle (this important class of non-straight lines which are equidistant to straight lines); or they lie on a horocycle, that is a circle of infinite radius.

[88] See for instance Proclus, *In Euclidis* 175–77, who endorses Euclid's statement about coplanarity. Cf. also the remark by Clavius, *Euclidis*, pp. 21–22, who sides with Proclus. The opponent to this addition to the definition of parallels was Simplicius (or rather Aganis, quoted by Simplicius), and thus a tradition of the *Elements* in the Middle Ages; see the remarks in Gerardo da Cremona's translation of an-Nayrīzī, in Curtze 1899 and Tummers 1994, p. 24; cf. Lo Bello 2003b, pp. 40–41.

[89] The coplanarity of intersecting lines is proved in *Elements* XI, 2, which is however normally considered a very weak demonstration (the property is nowadays normally assumed to be an axiom). In *Elements* XI, 7, Euclid proves that the line joining two parallels is in the same plane as them;

to equidistant lines, as it is clear that even though in a plane the equidistant line to a straight line may be straight (assuming the Parallel Postulate), it is not the case that every equidistant line to a straight line *in space* is itself straight. The most important example of a line equidistant to a straight line in space is the cylindrical helix around it. This was the example advanced by Geminus,[90] but it is a relevant case for Leibniz as well. The cylindrical helix is in fact a *uniform* line, following Leibniz' definition of uniformity; that is a line whose every part is congruent with another part (i. e. a line along which you may rigidly move a part of it). More than this, straight lines, circles and cylindrical helices (as Leibniz knows) exhaust the class of uniform lines.[91] Since Leibniz often bases his foundational efforts about the very concept of a straight line on its uniformity, it is important for him to realize that in a plane the uniform line equidistant to a straight line is straight, but in space this is no longer true. This means that proving the uniformity of an equidistant line to a straight line is not enough to prove the Parallel Postulate itself; we have seen, however, that Clavius and others relied on a (vague) notion of uniformity to ground

---

the proposition is of great foundational import, as in it Euclid explicitly assumes that two straight lines do not enclose a space, and probably from this passage in the demonstration of *Elements* XI, 7 someone added the spurious Sixth Postulate of Book One which states this property of straight lines. It is true, however, that in this *Elements* XI, 7, Euclid only relies on his definition of parallel lines to say that the parallels themselves are coplanar. As Euclid's definitions are probably just "nominal" in the Leibnizian sense, there is no mistake in assuming that parallel lines are coplanar: but then Euclid had to prove in *Elements* I, 31 (where he teaches how to construct a parallel line) that the construction is in a plane; but this is not explicitly done, as in Book One Euclid is doing plane geometry (there may be some tension between the general definition of parallels, which was conceived by looking at the stereometric books, and the planar treatment of parallels in Book One).

[90] See Proclus, *In Euclidis* 176.

[91] The debate on *homeomeric* curves, as Leibnizian uniform lines were called in Antiquity, started already in Ancient Greece (see Proclus, *In Euclidis* 104–107), but was highly important in the Early Modern Age, and was discussed by Galileo (*Dialogo*, Day One, in *Le opere*, vol. 7, p. 40), Hobbes (*De corpore*, xiv, § 3, in *Opera*, vol. 1, p. 156) and Descartes among others. On the topic in Antiquity, see Acerbi 2010; on the foundational relevance for Descartes, see Arana&Mancosu 2010. Leibniz was aware of the relevance of the helix since his first studies on *analysis situs* (see a fragment from 1679 in Echeverría 1995, p. 272) and discusses it in the late *In Euclidis* πρῶτα (GM v, p. 199). In the important essay *Uniformis locus*, which we have already quoted, he shows that uniformity is a property of several lines and surfaces, including the cylindrical helix, while one needs a stronger property (grounded on self-similarity, rather than self-congruence) to characterize the "flat" spaces, such as the straight line, the plane and the "absolute space". Here again the quote: "Uniformis locus seu sibi congruus, dici potest, cujus partes congruenter terminatae sunt congruentes. At locus sibi similis est, cujus partes similiter terminantes sunt similes. Loca sibi similia sunt non alia quàm recta, planum, spatium ipsum. Loca uniformia sunt omnia loca sibi similia, et praetera alia, nempe ex lineis quidem arcus circuli et helix cylindrica, ex superficies verò sphaerica et cylindrica ..." (in De Risi 2007, p. 582). To prove the Parallel Postulate Leibniz has thus to demonstrate that the equidistant line to a straight line is a self-similar, not just a uniform, line. Remember, finally, that the cylindrical helix was employed by Leibniz as a doubtful example of a non-planar uniform line in the *Discours de métaphysique* (see above, *note* 23 in § 3). From a modern point of view, homeomeric lines are curves with constant curvature and torsion (and these are the straight line, the circle and the helix). Since curvature and torsion together form a set of invariants which is sufficient to characterize a curve up to isometries, uniform curves (i. e. curves whose parts can be rigidly moved upon themselves) coincide with homeomeric curves.

the Postulate.[92] In any case, in our text Leibniz assumes the truth of the Postulate and would like to prove the equivalence between Euclid's definition and the definition of parallels as equidistant lines, and in particular that an equidistant straight line is parallel in Euclid's terms; this amounts to proving that an equidistant straight line is (a) non-intersecting (but this is obvious), and (b) coplanar to the straight line. Leibniz thus gives a long (and not completely faultless) proof of the coplanarity of two straight equidistant lines (the bulk of *text 31*). The demonstration, in particular, proves that given a straight line and a point external to it (in space), there is only one straight line which passes through the point and is equidistant to the given line; the issue of coplanarity proper is only touched upon in the proof. It may be interesting to remark that in discussing the uniqueness (in a plane) of the parallel line through an external point (which is a statement equivalent to the Parallel Postulate), Leibniz explicitly sees that it follows from the transitivity of the relation of parallelism (and this transitivity is in fact provable for equidistant lines). Proofs of the Parallel Postulate through the transitivity of parallelism (taken as non-incidence) were attempted in the seventeenth century, and one of them had been produced by Mercator, a geometer that Leibniz knew and appreciated.[93] In the end, Leibniz may claim to have proved that an equidistant straight line is also

[92] Clavius' discussion of the flow of an equidistant line never refers to the plane, and Leibniz added a handwritten "in eodem plano" in his edition of Clavius' book. In the context of tridimensional non-Euclidean spaces, the issue of equidistance becomes relevant. Although geodesics are not equidistant to other geodesics in hyperbolic geometry and plane elliptic geometry, in fact, tridimensional elliptic space admits geodesics which are equidistant, but not co-planar, to other geodesics. These are called *Clifford parallels* and have several relevant properties; for instance, one can recover Euclidean theorems about skew (non planar) Saccheri quadrilaterals formed by Clifford parallels (these behave in fact like Euclidean "rectangles" with four right angles and equal opposite sides – even though they are not in a plane), as well as some interesting connections with the cylindrical helix. Leibniz could not possibly be aware of these developments, however.

[93] A hint in this direction came from Proclus himself (see above, *note* 66), and we have seen that transitivity was discussed by Roberval. The most relevant attempt to ground the Parallel Postulate on the transitivity of parallelism in the seventeenth century, however, was made by Nikolaus Kauffmann (Mercator), in his *Euclidis Elementa Geometrica novo Ordine ac Methodo fere demonstrata*, from 1678. Mercator began by defining parallel lines as straight lines that are not 'inclined' toward one another, though without specifying the meaning of 'inclination': "Parallelae lineae sunt, quae non inclinantur ad se mutuo" (Definition 11; p. 2). He proceeded to take on an axiom according to which two lines are not inclined to one another if they are not inclined to a third: "Duo lineae non inclinantur ad se mutuo: quando earum una non magis quàm altera versus eandem partem inclinatur ad aliquam tertiam", and explains: "Sicuti duo homines, numquam assequentur sese mutuo, quando eorum unus, non magis celeriter quam alter, per eandem viam currit ad eundem terminum" (Axiom 3; p. 2). Next, he easily deduces (Theorem 7; p. 7) the transitivity of parallelism (*Elements* I, 30), then *Elements* I, 29 and the Fifth Postulate. We have already mentioned (see above *note* 34 in § 3) that Mercator was among those (few) geometers in the seventeenth century that defined a *point* (distinguishing it from a simple *minimum* in magnitude) through the notion of *position*, thus already pointing to a geometry of spatial relations. Leibniz was especially interested in Mercator's work on the quadrature of the hyperbola, which he quotes innumerable times, but he also knew and appreciated his edition of Euclid: "Mons. Mercator, un des plus habiles Geometres du temps a aussi donné des *Elemens de Geometrie*, où il fait voir par quelques Essais comment on pourroit joindre dans la Geometrie la clarté à la certitude" (*Projet et essais pour avancer l'art d'inventer*, in A VI, 4A, n. 205, pp. 968–69).

a parallel in Euclidean terms, both in a plane and in tridimensional space. He has to prove the inverse implication, that a non-intersecting line is equidistant, or that an equidistant line is straight; this amounts to proving "axiom 13", a task which however Leibniz now leaves unaccomplished.

Several attempts to prove the Parallel Postulate are to be found, finally, in Leibniz' longest work discussing Clavius' edition of Euclid, the all-important *In Euclidis* πρῶτα from 1712 (*text 36*).[94] This essay, which is probably Leibniz' masterwork in the epistemology of mathematics and the foundations of geometry, engages the theory of parallelism in sixteen different paragraphs on the definition of parallel lines. The discussion begins (§§ 1–2) by criticizing Euclid's definition of parallels as non-incident lines, the reality of which needs to be demonstrated, and in fact can be proved only by resorting to a deeper consideration (on equidistance, probably, or perhaps on the properties of space itself); Euclid had to accept the Parallel Postulate, in fact, since he did not have a good definition of parallel lines.[95] In the following section, Leibniz resorts to his definition of parallels as straight lines that *se invicem ubique habent eodem modo* (§ 3) and tries to claim that the flux of a point which keeps the same situation to a given straight lines (is equidistant to it) produces another straight line. The possibility of such a construction should follow from the uniformity of space and the uniformity of the straight line.[96] This line of reasoning surely improves on Clavius' own recourse to uniformity, as it starts from a well-defined concept of uniformity and the situational characterizations of space (as a uniform structure, which is isotropic or self-congruent), straight line (*utrinque se habens eodem modo*, the self-congruent line) and parallel lines (*se invicem ubique habentes eodem modo*). The appeal to a property of space, in particular, is especially new and pregnant with the future. This complex device of symmetries and uniformities is however still inadequate to prove the Parallel Postulate, and Leibniz realizes, on other occasions, that this is the case. In fact, he was able to distinguish (and it wasn't easy at the time) uniformity from self-similarity, ascribing the former to a plane and also (for instance) to a spherical surface, and the latter to flat (Euclidean) spaces and subspaces.[97] The uniformity of space is not a property strong enough to

[94] As Leibniz left it without date, the exact time of the composition of the essay is debated, but it seems to be very late. For an argument for 1712, see De Risi 2007, pp. 117–18.

[95] On Euclid's construction of parallels, see above *note* 29 in § 3.

[96] Note that in this § 3, Leibniz recognizes two motions of a point that preserve its situation to a straight line: the rotation of the point around the straight as an axis (generating a circle), and the flow of it equidistant to the line (generating a straight line equidistant to it). This double uniformity was already remarked by Leibniz in his previous attempts to define, *at the same time*, a straight line and a circle through the rotation of a body around an axis. See the already mentioned *Generatio quidem rectae et circuli*, quoted in *note* 42 in § 3; or Leibniz' (still unpublished) marginal notes to Barrow's *Euclidis* in *note* 34.

[97] The most relevant text is the above-mentioned *Uniformis locus*: see *note* 91. But in the same *In Euclidis* πρῶτα Leibniz also states similar ideas: "Plana intus uniformia sunt, nec nisi extremis distinguuntur. Illud solius planae superficiei proprium non est, ut congrua sint, quorum extrema sunt congrua, nam et superficiei sphaericae et cylindricae partes congruis circumferentiis inclusae congruae sunt" (GM v, p. 189). We may note that Proclus (*In Euclidis* 120) denies that a cylinder is a uniform (homeomeric) surface, as he doesn't see how any part of it can be fitted on the surface;

guarantee that the flow of a point equidistant to a straight line will produce another straight line. Leibniz has to go further. In the following sections (§§ 4–5) Leibniz tries one more time to employ his new definition of parallels as lines that have everywhere the same situation with respect to one another in order to further his aim. Leibniz constructs two points equidistant to a straight line and then joins them with the straight which is determined by them, claiming that it is an equidistant line. Even though he is right in pointing out the uniqueness of a straight line between two points (since it is not an affair concerning the Parallel Postulate, and he claims to have proved this principle elsewhere), it doesn't follow that a straight line which is equidistant to another straight line in two points is also equidistant to it in all other points. Giordano had proved, in fact, that three equidistant points are enough to infer the equidistance of all of them (and thus the Parallel Postulate); two points, however, are not enough, and Leibniz' proof falls short of a complete demonstration.[98] Then (§§ 6–7) Leibniz discusses again the issue of the parallel motion of a segment to draw a parallel line, and states that one has to prove that the equidistant line is straight. He mentions again the notion of uniformity, stating that straight lines are such; but this is insufficient, too, as he doesn't prove that an equidistant line is uniform (a hypercycle is not uniform in Leibniz' terms), nor that a uniform line is straight (it has to be self-similar to be straight, and self-similarity is a stronger property than uniformity). Leibniz then moves (§§ 8–9) to the definition of parallels as equidistant lines, which is (as we know) just an easy derivation from that of an identical reciprocal situation. He repeats that it is important to have a good definition of equidistance, and that this notion makes it possible to give a general concept of parallelism that also applies to curves and surfaces. Then (§§ 10–12)[99] he makes an attempt using Borelli's definition of parallel lines as lines that make equal angles with a transversal, and he correctly states that this definition is not of any help in proving the Postulate, as one has now to prove that lines that make equal angles with a transversal also make equal angles with *every* transversal (a statement which is equivalent to the Postulate itself); this is the reason why Borelli had eventually accepted the Postulate as an unproved principle. Leibniz tries to prove the latter statement about every transversal, but is clearly dissatisfied with his efforts; the manuscript bears traces of six more paragraphs of demonstrations of the Postulate, all of them erased by Leibniz himself. In the following lines (the actual § 12), moreover, the clear logical claim

think, for instance, of rotating a part before trying to place it back on the cylinder. Proclus (in our terms) was considering the properties of the embedding of a cylinder in three-dimensional space and the isometry group of the latter. Leibniz' more advanced stance relies on intrinsic properties of the (flat) metric of the surface – even though, of course, he didn't arrive at the Gaussian notion of intrinsic curvature.

[98] For Giordano's proof, see his (correct) Lemma 6 after Proposition 26 of Book One (*Euclide*, pp. 51–52). The same statement, that if two straight lines are equidistant in two points then they are equidistant also in a third point, was taken by D'Alembert as a substitute for the Parallel Postulate, while he lamented that such a simple proposition was still without proof (*Encyclopedie*, entry "*Parallèle*"; vol. 11, p. 906). On this lemma by Giordano, see Bonola 1905.

[99] In *text 36*, I have preserved Leibniz' original numbering of the sections, that Gerhardt had normalized in his edition. There is thus some discrepancy between our numbers and those in Gm v.

that a new definition of parallels cannot dispense with a proof of the Postulate (as the same difficulty will surface again in another form) is blurred by a confusion on the statement of the Postulate itself; Leibniz says that the axiom would amount to stating that Borelli's parallels are non-concurrent, but this is in fact easily provable without further assumptions (*Elements* I, 27). In the end, Leibniz admits that he has attempted much but not yet proved the Parallel Postulate (*multa tentavi...*).

And however: "... it seems to me that the whole thing may be obtained by a deeper principle, that of determinant reason". In § 13, Leibniz gives his proof of the Parallel Postulate through the principle of reason. He abandons the previous discussions on uniformity, equidistance and situation, and tackles the problem dealing with the interior angle sum of a quadrilateral.[100] He considers a quadrilateral with two equal sides at right angles from the base, and says that by the principle of reason the two superior angles cannot be different from one another, as everything is symmetric and nothing could determine a difference among them. The statement is true, and in fact it is provable without resorting to the principle of reason; it was proved already in the Middle Ages, and again by Giordano.[101] But then Leibniz adds that these two angles are not only equal to one another, but also right angles: since *there is no principle to determine their inclination*. This is equivalent to the Parallel Postulate. The principle of determinant reason is here applied to prove the Postulate through the idea that if there is no positive reason why the superior angles should be different to the base angles, then they are equal. It states a property of symmetry and uniformity (in a general sense) of absolute space, following the Leibnizian principle that every determination should have a positive cause.[102]

---

[100] On Saccheri quadrilaterals, see above the *note* 22 in § 2. Note that in § 12 Leibniz also mentions Lambert quadrilaterals, that is quadrilaterals with three right angles, whose fourth angle need to be determined.

[101] It is also Proposition 1 of Saccheri's *Euclides vindicatus*.

[102] There is a couple of other passages in which Leibniz uses the Principle of Reason to determine a geometrical figure. In these cases, however, the line of reasoning is not purely geometrical, but concerns the choice of God. Leibniz claims that God, if He had to create a triangle and had no further reasons to shape it in one way or the other, He would produce an equilateral one; similarly, if He had to connect a point with a straight line, He would do this employing a perpendicular. These quite abstract (and popular) examples go in a different direction than the present proof of the Parallel Postulate, as they assume that symmetry is a perfection, and thus God may find a *reason* to act in symmetry itself. Symmetry thus becomes a motive for action, and substantiates what Leibniz calls the Principle of Perfection (or the Principle of the Best), which only applies to the actual world (the best of all possible worlds). On the contrary, in the present context symmetry is regarded as a feature of space itself, with no reference to the choice of God nor the existence of the actual world. The proof by symmetry of the Parallel Postulate relies on the Principle of Reason itself, that seems to have a different epistemological status than the Principle of Perfection. In particular, the latter depends on the will of God, while the former (like the Principle of Contradiction itself) seems to be independent of it. The Principle of the Best is clearly contingent; that of Determinant Reason seems to be necessary and still not analytical. Here are the quotes by Leibniz: "Ce principe de la nature d'agir par les voies les plus determinées que nous venons d'employer, n'est qu'architectonique en effect, cependant elle ne manque jamais de l'observer. Supposons le cas que la nature fut obligée generalement de construire un triangle, et que pour cet effect la seule peripherie ou somme de costés fut donnée et rien de plus, elle construiroit un triangle equilateral.

Leibniz has only one other geometrical proof in the *In Euclidis* πρῶτα (and in his whole scientific production) that works through the Principle of Reason: that two straight lines only meet in one point. Leibniz' line of thought, in this case, is that there is *no reason* to establish the distance at which two incident straight lines should meet again; thus, they simply cannot meet again. Leibniz considers a pencil of straight lines radiating from point A, and assumes that two of them (α and β) meet again in another point B. Then he considers the distance between A and B. Such a distance seems to him to be evidently determined by the angle comprised between the two straight lines – the greater the angle, the longer the segment AB.

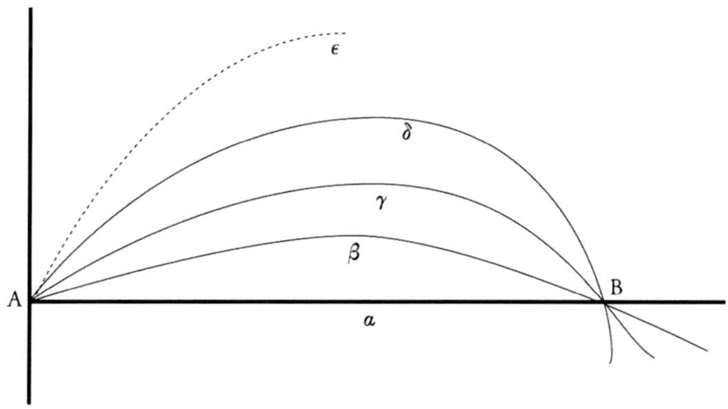

Given the definition of a straight line as a line that cuts the plane into congruent parts, it will certainly be possible to take a third straight line in the pencil starting from A that also ends in B. In fact, as the straight line AβB divides the plane into congruent parts, in much the same way as it has the straight line AαB on its right, it can well have its counterpart AγB on its left. Repeating the procedure for each straight line so obtained, it will be possible to demonstrate that infinite straight lines AδB, AεB, etc. starting from A and ending in B, exist. If it is so, however, the distance between A and B will in no way depend on the angle comprised between the straight lines, as it always remains the same for all infinite straight lines of the pencil. But since the reciprocal situation of α and β is given only by the angle they form,

On voit par cet exemple la difference qu'il y a entre les determinations Architectoniques et les Geometriques. Les determinations Geometriques importent une necessité absolue, dont le contraire implique contradiction, mais les Architectoniques n'importent qu'une necessité de choix, dont le contraire importe imperfection. … Si la nature estoit brute, pour ainsi dire, c'est à dire purement materielle ou Geometrique, le cas susdit seroit impossible, et à moins que d'avoir quelque chose de plus determinant que la seule peripherie, elle ne produiroit point de triangle; mais puisqu'elle est gouverné Architectoniquement, des demy-determinations geometriques luy suffisent pour achever son ouvrage, autrement elle auroit esté arrestée le plus souvent" (*Tentamen anagogicum*, GP VII, pp. 278–79); and "Ce seroit autre chose, si Dieu decernoit de tirer d'un point donné une ligne droite jusqu'à une autre ligne droite donnée, sans qu'il eût aucune determination de l'angle, ny dans le decret, ny dans ses circonstances; car en ce cas, la determination viendroit de la nature de la chose, la ligne seroit perpendiculaire, et l'angle seroit droit, puisqu'il n'y a que cela qui soit determine, et qui se distingue" (*Theodicée*, § 196; GP VI, pp. 232–33).

Leibniz says, the distance between A and B will depend on nothing at all. There is *no reason whatsoever* that may determine the distance AB. But this is impossible, and thus there is only one straight line through AB.[103]

Leibniz is correct in saying that the distance AB (the injectivity radius at A, in modern terms) does not simply depend on the angle at A. It depends, in fact, on the curvature radius of the manifold. Leibniz is thus claiming that there is no *a priori* reason why this radius should have one positive value or another.

The same is true with respect to the upper angles of the Saccheri quadrilateral that we were considering in Leibniz proof of the Parallel Postulate. The size of these angles is dependent on the space curvature, and if the curvature is positive then the angles will be obtuse, while if the curvature is negative they will be acute. But then Leibniz would say that there is no principle to determine the exact positive or negative curvature of the manifold, so it has to be zero. In other words, since there is only one curvature value (zero) that gives a Euclidean space, while there are plenty of different elliptic and hyperbolic spaces (with any positive or negative curvature value), and there is no reason to select one value from another among these classes, space has to be Euclidean.[104] The Principle of Reason may be applied because there are qualitative differences in the behavior of parallels in the three isotropic (uniform) geometries: for instance, in Euclidean geometry only a parallel to a straight line passes through a point external to it, while in hyperbolic geometry an infinity of parallels pass through the point, and none in elliptic geometry. This is why, in fact, we do speak about *three* geometries. The space curvature, however, is a continuous function, and it is only exactly determined (as zero) in the Euclidean case. We have to stress again that the lack of reason for an *a priori* determination of an exact positive or negative curvature of space may become in Leibniz' thought a positive reason for flatness since he considers a curvature of zero to be less determined than any other given value of it.

It is highly remarkable, in any case, that the only two instances of the employment of the Principle of Reason in geometry ground, respectively, the impossibility of an elliptic space and that of a hyperbolic space. For a modern reader it is almost impossible to avoid the impression that the simple analysis of the concept of space would be enough, for Leibniz, to determine all the features of it (including isotropy), with the sole exception of the exact curvature that makes it Euclidean, elliptic or hyperbolic; and a further principle is needed to choose among the three.

---

[103] This is in § 5 of Leibniz' proof of Clavius' Axiom 14 in the *In Euclidis* πρῶτα (GM v, p. 210), that, given its importance for the present discussion, I have reproduced as the final part of *text 36*. A similar proof is given by Leibniz in § 2 (GM v, p. 209–10), based on the fact that it is certainly possible (by the above-mentioned reasoning, and applying the axiom of Archimedes) to find a straight line AωB that makes an angle with AαB greater than the right angle. But this is absurd, since Leibniz claims that two perpendicular straight lines meet in one and only one point. I am drawing here from my discussion on the topic in De Risi 2007, pp. 252–55.

[104] A similar line of reasoning is endorsed by Leibniz even in discussing God's choice of the best possible world: God chooses, in fact, the most perfect world that is unique as regards its degree of perfection. If there were two most perfect worlds, God would have had to choose to create a less perfect universe, as He could not have any reason to pick the one or the other among the most perfect ones. See a hint in this direction in § 73 of the *Generales inquisitiones* (A vi, 4a, n. 165, p. 763).

Leaving aside these last considerations, which seriously risk anachronism, we could remark that grounding a general theory of parallelism and incidence on the Principle of Reason may save from a contradiction the useful fictions of projective geometry and similar mathematical tools. We have already seen (*text 11*) that Leibniz distinguishes between properly impossible things (that entail a contradiction) and imaginary things (*seu impossibiles per accidens*), and praises the use of these in mathematical reasoning. His examples are the infinitesimals, imaginary numbers, and, most of all, the use of points of incidence at infinity in the projective treatment of conic sections. It is thus at least possible that Leibniz could see in his late discussion of parallels through the Principle of Reason an important logical ground to conceive projective techniques as consistent and useful fictions. Perspectival geometry, which is the only "non-Euclidean" geometry that Leibniz could historically conceive, is not strictly true, since projective space is an imaginary thing that does not obey the Principle of Determinant Reason and thus cannot exist; but it is not inconsistent, and may be used to prove important theorems about real spatial structures, as well as to pictorially represent real space.

Leibniz' proof of the Parallel Postulate through the principle of reason is developed somewhat in the following paragraphs (§§ 14–16) of the *In Euclidis* πρῶτα, where Saccheri's quadrilaterals are used to prove the straightness of the equidistant line generated through the flow of the segment, as well as to prove in a simpler way *Elements* I, 27 and 28 (which in fact do not require the Parallel Postulate). This proof is not very relevant in itself, but it may have some historical significance as Proclus reports a demonstration of *Elements* I, 28 by Ptolemy, in which the ancient mathematician employs a principle of symmetry that can reasonably be connected with the Principle of Reason. This is probably the only proof in Greek geometry that engages with similar considerations, and it is highly remarkable that it appeared in a treatise aimed at proving the Parallel Postulate. We have no hint, however, that Leibniz had Ptolemy in mind when he envisaged his own proof.[105] In the following lines Leibniz again mentions Proclus' and Clavius' proofs of the Parallel Postulate. He then con-

[105] Proclus informs us that Ptolemy's proof was part of an essay on the Parallel Postulate. The theorem of *Elements* I, 28 says that "if a straight line falling on two straight lines makes the interior angles on the same side equal to two right angles, the straight lines will be parallel to one another". Ptolemy's proof is that if the lines met on the one side, they would also meet on the other side, because the equality of the angles on the two sides makes them indistinguishable and there is no reason why the lines should meet only here or there. But this is impossible as two straight lines cannot enclose space (Proclus, *In Euclidis* 362–63). The latter is the main principle used in the proof, of course, as in spherical geometry lines meet in two points and *Elements* I, 28 is false. We don't know exactly how Leibniz' proof based on the Principle of Reason worked. In § 16, he gives a different one, but also mentions the possibility of constructing a proof based on the results proved at § 13 (i. e. the demonstration of the Parallel Postulate through the Principle of Reason). We may remark that Claude Richard in his edition of the *Elements* had explicitly assumed as a principle (second part of Axiom 13, p. 14) that if two straight lines make equal internal angles with a transversal on one side, and meet on that side, they should also meet in any other case in which they form equal internal angles with a transversal (and thus, on the other side as well). This is a kind of formalization of Ptolemy's symmetry principle. In the eighteenth century, Andreas Segner resuscitated Ptolemy's proof of *Elements* I, 28, in his *Vorlesungen über die Rechenkunst und Geometrie* from 1747 (where he also employed the notion of *Lage*, that is *situs*: see below *note* 18 in § 5).

cludes (§ 17) with a summary of the various definitions and the main difficulties in the demonstration. Further on in the essay, Leibniz briefly discusses again the topic when his commentary arrives at the Parallel Postulate proper (Clavius' Axiom 13). In this connection, however, Leibniz only rephrases Euclid's postulate, expressing it as stating that "if two straight lines don't make equal angles with another straight, than they converge". This is simply a condition of *Elements* I, 29, which is in fact equivalent to the Postulate but which Leibniz finds more evident than the original; and Leibniz has deleted the part stating that the two straights meet "on the side on which are the angles less than the two right angles", as it is easily provable once one assumes (or proves) that they meet somewhere. Leibniz promises to give a full proof of the Parallel Postulate in the following pages, when he comes to deal (as Clavius before him) with *Elements* I, 29; the *In Euclidis* πρῶτα, however, he only discusses some further axioms and then breaks off.

Reception and Legacy 5

5.1 Toward a Geometry of Space in the Eighteenth Century.

Leibniz did not publish anything of his new geometry, the *analysis situs*, and nothing of his researches on the Parallel Postulate. His influence on the subsequent developments of Euclidean and non-Euclidean geometry was almost inexistent. Yet, we may find a few episodes in the history of mathematics that show at least an indirect (and perhaps misconceived) reception of Leibniz' endeavors.

The very fact that Leibniz attempted during all his life to build a new geometry emerged as early as the eighteenth century. Many of his correspondents knew about his efforts, even though they may have not seen a single result of these researches. Already in 1713, Christian Wolff (1679–1754) declared to his many readers that the great Leibniz was probing the depths of geometry through a new *analysis situs* grounded on the concept of geometrical similarity; he did not know anything else on the subject, but dared some contributions in the same direction (whatever it was). The edition of Johann Bernoulli's correspondence in 1745, showed that Leibniz did indeed work on a new geometry, but revealed nothing about its contents. In 1753, the Leibnizian physicist Samuel König (1712–1757) edited a letter of Leibniz (addressed to Huygens), in which the German philosopher attempted to convince his former teacher of the advantages of a *characteristica geometrica*, giving a few samples of it.[1] The *name* of *analysis situs* began to become renowned in the field of geometrical

[1] Wolff's quotation comes from his *Elementa Matheseos universae*, vol. 1, p. 296: "Neque vero putandum est, integram Analysin jamdum esse inventam; [...] Certe, quae in elementis Geometriae docuimus, per modernam analysin non omnia eruuntur, inprimis si a linearum & superficierum situ pendent. Quamobrem Leibnitius, vir in omni eruditione summus, pro ea, quae ipsi est, ingenii perspicacitate novam quendam analysin situs excogitaverit, peculiari calculi generi (quem calculum situs appellat) superstructam, a calculo magnitudinum, quibus in nostra Analysi utimur, toto coelo differentis". Leibniz mentioned his *analysis situs* several times to Bernoulli, who however never got interested enough in Leibniz' project; their correspondence, however, bears traces of Leibniz' endeavor. Something more is to be found in his letter to Huygens from September 8th,

V. De Risi, *Leibniz on the Parallel Postulate and the Foundations of Geometry*,
Science Networks. Historical Studies, DOI 10.1007/978-3-319-19863-7_5

studies as one of the most alluring intellectual enterprises, which, had it been fully accomplished, would have renewed the entire field of mathematics. Euler presented his own study on the Bridges of Königsberg (which is now considered the starting point of algebraic topology) as an example of *analysis situs*; Buffon hoped that its application to zoology would have brought important results; Lambert wanted to apply it to building a new mathematical logic; Kant regarded it with some skepticism.[2] None of them had the faintest idea of what Leibniz had attempted to do.

They were aware, however, of Leibniz' epistemological views about geometry, as a few of his philosophical papers had been published during his lifetime or shortly thereafter. Leibniz' epistemology had a wide impact on eighteenth-century philosophy and mathematics, and the whole controversy on the mathematical method that was disputed in Germany for fifty years was mostly fought in Leibniz' name or against him. Many different issues were at stake, the meaning and relevance of nominal and real definitions, the role of constructions, the necessity of diagrams in geometry, the exact extent and use of conceptual analysis in geometrical reasoning, the possibility of proving the axioms and the distinction between axioms and postulates, the need (or not) for a new symbolism and the relations between classical geometry, algebra and *characteristica* in general, and of course the connection of the mathematical method with that of philosophy (and much more). In the background of these discussions, we may see the slow transformation of geometry from a discipline dealing with figures and magnitudes into a science of space: a new definition of geometry that no one endorsed at the beginning of the century (with the exception of Leibniz) but that would become widespread and almost universally accepted in the first decades of the nineteenth century. Most of these developments in epistemology were, in fact, intertwined with the attempts to prove the Parallel Postulate, which eventually ended up in the recognition of the possibility of non-Euclidean geometries. As soon as space was generally understood and conceived as a mathematical structure, a plurality of such structures were admitted and investigated. In this respect, the history of geometry and geometrical epistemology in the eighteenth century had totally un-Leibnizian outcomes (the end of geometrical "logicism" and the birth of non-Euclidean geometries) framed in a general understanding of geometry as a science of space that Leibniz was the first to envisage and foster.

1679 (now in A III, 2, n. 359, pp. 888–89), which was published almost by chance by König in his *Appel au publique*, a polemical paper written against Maupertuis during their controversy on the discovery and meaning of the Principle of Least Action in mechanics. On this quarrel, which inflamed Europe for a couple of years in the middle of the eighteenth century, and involved the name and reputation of Leibniz, see Barber 1955, Pulte 1989, Beeson 1992, Goldenbaum 2004 (pp. 508–651).

[2] These are just a few examples, of course. Euler's paper is the *Solutio problematis ad geometriam situs pertinentis*, presented to the Academy of Sciences in 1735 and published in 1751 (now in *Opera omnia*, I, 7). Buffon mentioned Leibniz' *analysis situs* in chapter 9 of his *Histoire des Animaux* from 1750 (*Oeuvres complètes*, vol. 4, p. 663). Lambert did the same in the logical essay *De universaliori calculi idea* from 1765. Kant' opinion is famously stated in his *Von dem ersten Grunde des Unterschiedes der Gegenden im Raume* from 1768 (KgS II, p. 377).

The story of Lambert's attempt to prove the Parallel Postulate is very telling. For many years, he fought against Leibniz' conception of geometry (which he knew in a very mutilated form), complaining about the modern practice of assuming space itself as an object of geometrical inquiry, and stating that Euclid was right in excluding any reference to it in his *Elements*. And yet, after having attempted, and failed, to prove the Parallel Postulate, Lambert was forced to assume it as a first principle of geometry. Dealing with the theory of parallels, however, he came to realize (as Leibniz before him) that the Parallel Postulate cannot be regarded as a simple principle about the behavior of certain straight lines and angles, and that it captures a deeper structure that innervates space itself. In his last philosophical work, Lambert accepted Wallis' principle on the possibility of transformations by similarity as expressing the veritable meaning of the old Postulate, and stated it as a principle *about space* (rather than parallels, straight lines or angles). This statement conflicted with Lambert's overall epistemology, and it may appear to be a surrender to Leibniz' position. It is highly remarkable, in any case, that the investigations on the Parallel Postulate had pushed Lambert, almost against his will, to this Leibnizian conclusion; a conclusion (the consideration space as an object of enquiry), that was the first precondition for the development of alternative geometries.[3]

5.2 Wolff, Kästner and the Early Reactions.

The attempts to prove the Parallel Postulate were very numerous in the eighteenth century, and they were especially concentrated in Germany as they were connected with the general epistemological debate on the nature of geometry that was going on there. The best book on the topic, however, was Saccheri's *Euclides vindicatus* from 1733; it had almost no circulation in the century, but was aptly epitomized in a 1763 booklet by Simon Klügel (1739–1812) that listed several failed attempts to prove the Parallel Postulate.[4] Klügel's *Recensio* enjoyed widespread dissemination,

[3] Lambert's statements about the impossibility of doing geometry by an analysis of the concept of space should be framed in his general polemics against Wolff's and Leibniz' theory of definition. For a few quotes in this direction, see for instance Lambert's letter to Holland from April 21st, 1765: "Euclid hat keine Definition von der Geometrie, und muß allem Ansehen nach gar nicht daran gedacht haben, diese Wissenschaft aus der Definition derselben herzuleiten. … Indem Euclid sich sogleich zu den Linien, Winkeln, Triangeln etc. wendet, so nimmt er nicht die Analyse sondern die Anatomie des Raumes vor und dadurch bringt er die Geometrie zu Stande" (Lambert, *Briefwechsel*, vol. 1, pp. 31–32); and the letter to Kant from February 3rd, 1766: "Euclid leitet seine Elemente weder aus der Definition des Raumes noch aus der Definition der Geometrie her, sondern er fängt bey Linien, Winkeln etc. als dem einfachen in den Dimensionen des Raumes an" (p. 350). Note that the letters are from 1765 and beginning 1766, while Lambert would compose his *Theorie der Parallellinien*, and realize the difficulty (or impossibility) of proving the Parallel Postulate in mid-1766. In § 79 of his *Anlage zur Architectonic* from 1671, he states Wallis' principle about *space* (vol. 1, p. 62).

[4] Klügel's work was titled *Conatuum praecipuorum theoriam parallelarum demonstrandi recensio*, and presented an abridgement of several attempts to prove the Parallel Postulate; it was the only source of Saccheri's circulation in Germany, as the *Euclides vindicatus* was not read by

and through it German mathematicians became especially interested in the topic; in the fifty years preceding Bolyai's and Lobachevsky's discoveries (around 1830) dozens of books and hundreds of papers were written in a desperate attempt to prove the Parallel Postulate.

Some of these attempts bear the trace of Leibniz in several respects. Wolff's own opinion on the Parallel Postulate was that it could be demonstrated starting from the definition of parallels as equidistant lines, which he accepted without realizing that it would need a justification.[5] Wolff agreed, moreover, that any geometrical definition had to be constructively justified to be "real"; but apparently he thought that the definition through equidistance was constructive enough to rely on it.[6] As we see, he completely lacked an analysis of situation and the idea that even the possibility of a construction in space has to be justified through the properties of space itself. Notwithstanding a nominal endorsement of Leibniz' view that geometry is the science of space, Wolff's geometry is that of Euclid: and there is no definition, or axiom, or theorem, in all the volumes that Wolff devoted to geometry, that mentions space or its properties. He knew the names of *analysis situs* and *scientia spatii*, but not the things themselves. He wanted, in any case, to pursue Leibniz'

anyone in the original. Klügel's work, on the contrary, was read and quoted by every mathematician in the country, and many of them got interested in the theory of parallels through it (Lambert, for one, wrote his *Theorie der Parallellinien* mostly inspired by Klügel's book). An English translation of Klügel's work will be available in my forthcoming edition of Lambert's *Theory of Parallel Lines*.

5 Wolff's definition of parallel lines as equidistant straight lines is in § 81 of the *Elementa geometriae*, in *Elementa matheseos universae* (vol. 1, p. 103). His proof of the Parallel Postulate is in the following § 262 (p. 130), based on a strategic application of the definition of equidistance in § 257 (p. 128). In the German *Anfangs-Gründe aller mathematischen Wissenschafften*, Wolff defines parallel lines as equidistant lines (vol. 1, p. 129; ed. 1738) and skips the Postulate to directly prove *Elements* I, 29 (pp. 191–92). Wolff's mistake is especially egregious since he shows knowledge of Saccheri's work (he quotes *Euclides vindicatus* in *Elementa matheseos universae*, vol. 5, p. 28), and Saccheri argues for pages and pages about the inadequacy of the definition of parallel lines through equidistance.

6 In his philosophical works, Wolff insists on the necessity of real and constructive definitions, and claims that the Euclidean definition of a circle should be improved into the one given through the rotational flow of a segment: see his German Logic (*Vernünfftige Gedancken von den Kräften des menschlichen Verstand*, I, § 51, p. 148; ed. 1727). In the Latin Logic Wolff restates the same idea and quotes Leibniz' *Meditationes de cognitione, veritate et ideis*, the only published text in which Leibniz traced the distinction between nominal and constructive definitions (without giving the example of the circle): see Wolff, *Logica*, § 191 and 194–196 (pp. 211–14). Wolff's *Anfangs-Gründe aller mathematischen Wissenschafften* (first published in 1710) also define the circle through rotation (vol. 1, p. 121), and the equidistant parallel lines through the perpendicular flow of a segment on a straight line, as in the usual Clavian construction (p. 129). In his Latin treatise on mathematics (from 1713), however, Wolff gives nominal definitions and then justifies them by proving theorems (or rather problems) that allow the construction of the defined objects – he follows, in fact, the Euclidean method. So, for instance, he gives a nominal definition of a circle in *Elementa geometriae*, § 37 (*Elementa matheseos universae*, vol. 1, p. 100) and shows how to construct one (easy: with a compass!) in § 131 (p. 108). He then constructs parallel lines employing the proof of *Elements* I, 31 in *Elementa geometriae*, § 258 (p. 128), without realizing that Euclid's demonstration is valid for non-intersecting lines, and he should prove that parallel lines so drawn are in fact equidistant (in hyperbolic geometry, they are not).

grand program of a complete proof of the geometrical axioms, and a reduction of them to the definitions.[7]

Wolff's clumsiness in dealing with the Parallel Postulate engendered a wide discussion on the risks of relying on definitions in proving the geometrical axioms. Several philosophers and mathematicians criticized Wolff's geometry and its epistemology, and it is not hard to understand that Leibniz' name was involved in these quarrels. Even though everyone recognized that Leibniz' stature as a mathematician was incomparable with Wolff's poor treatment of elementary geometry, their philosophical views were often conflated together, and in fact Wolff endorsed several theses of Leibniz' mathematical epistemology. Moreover, as we know, the only papers published by Leibniz that mentioned the theory of parallels (*texts 19*, *24* and *34*) define parallel lines (straight or not) as equidistant lines – as Wolff himself did.[8] When the *Nouveaux Essais* were published in 1765, one more passage (*text 33*) showed evidence of Leibniz' preference for a definition through equidistance. As a consequence, several authors who wanted to defend the Leibnizian-Wolffian philosophy, also defended the epistemological idea that axioms should be derived from real definitions, and that the definition of parallel lines as equidistant lines allows us to dispense with the Parallel Postulate. Among them, we may mention at least Friedrich

---

[7] The first paragraph of Wolff's essay on geometry in his *Anfangs-Gründe* states: "Die Geometrie ist eine Wissenschafft des Raumes" (vol. 1, p. 117), but then Wolff seems to forget this definition and goes on giving the usual Euclidean definitions of the basic mathematical figures (straight lines, triangles, etc.) and the theorems about them; space itself is not mentioned anymore. Wolff's opinion about axioms is that they are strictly unprovable, "propositiones theoreticae indemonstrabiles" (see his Latin *Logica*, §§ 267–68, pp. 258–59), since they are tautological with definitions: a geometrical axiom being a proposition like "all radii of a circle are equal", which immediately follows from the definition of a circle (*Logica*, § 262, p. 255; cf. Borelli's similar statement in *note* 27 in § 3). He admits that identical propositions are axioms, as Leibniz said (*Logica*, § 270, p. 259–60), but would like to extend them to a more comprehensive class of propositions that "immediately" follow from definitions, such as that "a triangle is not a square" (*Logica*, §§ 271–74, pp. 260–61). In this way, probably, Wolff would like to accommodate Leibniz' strictest notion of a proper axiom as an identity with the mathematical practice, and try to explain the commonly used notion of "immediateness" in this connection (see above, *note* 2 in § 3). In the German Logic, Wolff embraced a more Leibnizian doctrine: see *Vernünfftige Gedancken*, III, § 13 (pp. 161–62); and VI, §§ 1–2 (pp. 190–92). Leibniz wrote to Wolff about his own plan to prove all the axioms in the very first letter he sent to him: "Malo deducere ex definitionibus Axiomata, quam assumere" (*Briefwechsel mit Wolff*, p. 19). As we have seen, in any case, a similar epistemological program was endorsed by several mathematicians. A compact abridgement of Wolff's epistemology of mathematics may be found in his *De methodo mathematica*, in *Elementa matheseos universae* vol. 1, pp. 5–17. A longer discussion is in the verbose dissertation by the Wolffian scholar Gottlieb Hagen, which however adds nothing momentous to his master's opinion, except several classical and modern sources: see Hagen's *Mediationes philosophicae de methodo mathematica*, from 1734. It should also be noted that Wolff made no distinction between innate and acquired ideas, which is one of the cornerstones of Leibniz' philosophy of mathematics (see above, *note* 44 in § 3); this distinction was not available in the Leibnizian papers published at the time of Wolff. See TONELLI 1974.

[8] There is no evidence that Wolff took his definition of parallel lines as equidistant straights from Leibniz' papers on differential geometry; and he died before the publication of the *Nouveaux Essais*. The actual sources of Wolff's *Elementa matheseos universae* are numerous and difficult to establish with exactitude. We have already seen, however, how common the definition of parallels through equidistance was in the seventeenth century.

Behn (1734–1804), whose 1761 *Dissertatio* on the theory of parallels discussed at length the notion of *situs* and equates it with distance, thus proposing a (quite naïve) proof of the Parallel Postulate based on equidistance, which captures nonetheless some genuine Leibnizian insight.[9] On the other hand, it should not be forgotten that the most important treatise on the theory of parallels in Germany, Lambert's *Theorie der Parallellinien* (1766; published in 1786), was for the most part an essay against Wolffian (and Leibnizian) epistemology, that advanced the all-important claim that axioms are the very foundations of geometry, while definitions should *follow* from them (rather than the other way round). It was thanks to this inversion in epistemology that, eventually, non-Euclidean geometries were first conceived as possible outcomes of this field of research. Kant himself, who fashioned the most influential anti-Leibnizian theory of geometry and space, felt the need to attack Wolff's definition of equidistant lines and his proof of the Parallel Postulate.[10] The two different debates, the one on the Leibnizian epistemology, and the other on the theory of parallels, were indeed nothing but different sides of the same coin.

Therefore, it comes as no surprise that Leibniz' mysterious *analysis situs* and the elusive proof of the Parallel Postulate were often brought together. Abraham Kästner (1719–1800), the most influential mathematician in Germany in the second half of the century, maintained that the Parallel Postulate was probably unprovable.[11] His

---

[9] Behn's *Dissertatio mathematica sistens linearum parallelarum proprietates* begins with a philosophical discussion of the notion of *situs*, stating that it is a relation of remote coexistence which constitutes the distance between things: "Sed quid per *situm* sit intelligendum, si quispiam quaerat, uno ferme ore omnes philosophi responsum ferunt, per situm innui eam relationem, quam res ob remotam coexistentiam inter se habent. Qui igitur manet idem, donec remota illa coexistentia non est mutata. Remota ista coexistentia constituit *distantiam*" (pp. 7–8). His reference to "*omnes philosophi*" is mostly to the Wolffian school, which in fact discussed a lot about the metaphysical meaning of situation, often in relation with the difficult issue of whether monads are situated or not. Later in the essay, Behn defines parallels as equidistant lines, and even asks whether the line equidistant to a straight line has to be straight, answering however with a few weak examples (Scholium to § 5, pp. 9–10; cf. the criticisms by Klügel, *Recensio*, § 23, pp. 26–28).

[10] Neither Lambert nor Kant, of course, envisaged non-Euclidean outcomes of their research. On Lambert's epistemology of mathematics and its role in the *Theorie der Parallellinien*, see the *Introduction* to my forthcoming English edition of this book. On Kant's studies on parallels, see his above-mentioned *Reflexionen* 8–10 (KGS XIV, pp. 33–51), and WEBB 1987, PELLETIER 2009, DE RISI 2013, HEIS [B].

[11] Kästner informs us that he was convinced for a long time by a proof of the Parallel Postulate given by Christian Hausen (1693–1743) in his *Elementa matheseos* from 1734 (Proposition 13, pp. 101–102, grounded on a mistaken Scholium 4 to Proposition 10, p. 100), but then, realizing that Hausen had committed a simple mistake, he spent several years attempting to prove the Postulate himself, amassing at the same time a large collection of books dealing with the problem. In the end, he became skeptical on the issue, and offered Simon Klügel the use of his library to write the 1763 dissertation. This biographical sketch by Kästner is to be read in the *Vorrede* of his famous *Anfangsgründe*, where he accepts the Parallel Postulate as Axiom 9 (pp. 205–206), limiting himself to presenting a few unproved but intuitive statements that may convince the reader of its truth. We may note that Kästner's collection of books on the Parallel Postulate in the library of Göttingen was quite important, as it included Saccheri's *Euclides* and Giordano's *Euclide*, two works that were otherwise almost unknown at the time. There are many speculations about the possibility that Gauss, himself a student in Göttingen, may have had direct access to these volumes. Kästner was also very

argument, however, was very classical: we don't have a good definition of a straight line, thus we are unable to demonstrate anything so fundamental as an axiom on parallel straight lines. He strengthens this argument saying that probably we are not able to fashion such a definition in principle, and we are thus forced to accept unproved axioms in geometry.[12] Kästner pressed his pupil Klügel to write the above-mentioned dissertation on the failed attempts to prove the Parallel Postulate: a paper that was aimed at showing the futility of such an endeavor, and provoked, as a reaction, the greatest production ever of essays on parallel lines. While publicly commenting on his student's work, however, Kästner himself raised some expectations of a proof, saying that *perhaps* the solution of the problem will be found when we are able to reconstruct the lost *analysis situs* that had died with Leibniz.[13]

This attitude toward the *analysis situs* became quite common. Some authors, following Kästner's line of enquiry, tried to employ this lost science to fashion a good definition of a straight line,[14] while several Leibnizian scholars abandoned the notion of equidistance and claimed that a non-metric, "qualitative" geometry might be the key to solving the problems of parallelism.

5.3 The Concept of *Lage*, Direction, and the Actual Infinite.

We have already mentioned that a few definitions of parallels in the seventeenth century were grounded on the notion of the "inclination" of a straight line; in the eighteenth century, the same concept resounded with Leibnizian harmonics and evolved into a theory of the "position" or "situation" of the lines; later on, in the nineteenth century, it eventually became a theory about the common "direction" of parallel lines.[15] The concept of the inclination of a straight line, was mostly conceived as a

fond of Leibniz, and almost a Leibnizian himself; he wrote the Preface to the first edition (1765) of Leibniz' *Nouveaux Essais*, where (as we know) Leibniz seemed to endorse an equidistance-theory of parallel lines (in any case, Kästner did not mention the issue in his preface).

[12] For Kästner's opinion on the definition of a straight line, see above, *note* 1 in § 4. His doubts about the possibility of obtaining a definition of it are spelled out in his *Ueber die geometrischen Axiome*, §§ 23–24, p. 430. See also his *Anfangsgründe*, pp. 180–81.

[13] Kästner's remark to this effect was printed at the end of Klügel's *Recensio*: "Habituros nos aliquando, veram eam cuius admoto geometriae lumine spectra dissipasti demonstrationem, vix speraverim nisi diligentius exculta doctrina situs, cuius analysis cum Leibnitio interiit" (p. 33).

[14] We have already mentioned that several important mathematicians (Saccheri, Gauss, Lobachevsky) who dealt with the theory of parallels accepted the definition of a straight line as a rotational axis, which was widely discussed by Leibniz. The definition had some circulation in Germany, and for instance a *Commentatio mathematica* (1789) by Johann Voigt claims (§ 19, p. 10) that with this definition one is able to prove the Parallel Postulate (he gives a very faulty demonstration of it in the following § 43, pp. 23–24).

[15] Klügel himself, in § 25 (p. 29) of his *Recensio*, mentions a few authors (Clairaut, the aforementioned Sauveur, Camus, Boscovich) who had employed the notion of inclination to define parallel lines. Even though Klügel's references are not always correct, it is true that many mathematicians in eighteenth-century Germany attempted this path. I have suggested that Mercator is in fact the best example in this connection, and the one who has the most polished theory of parallelism as

derivative concept, and it was defined through some angular notion (as in Borelli). I would say that a veritable situational theory of parallels begins with the assumption that the concept of inclination needs no further analysis. This trend probably began with the *Theorie von den Parallellinien* (1778) by Wenceslaus Karsten (1732–1787), which called "parallel" those lines that have the same *Lage*. The German word "*Lage*" was normally used, at the time, to translate the Latin "*situs*", and even though Karsten did not explicitly mention Leibniz in this connection, it is not difficult to guess his source; he went on, in fact, to envisage a theory of position (a *Geometrie der Lage*) that was intended to complement the usual geometry of magnitude, thus proposing a research program which derived from Leibniz' *analysis situs*. In a further paper on the subject, he equated the *Lage* of parallel lines with their common *Richtung* (direction). This notwithstanding, Karsten's theory still refers to the angles that a pencil of parallels make with a transversal line, thus defining the sameness of *Lage* through the usual Euclidean property.[16] However, the concept of *Lage* and its inclusion in a broader mathematical program of a geometry of position dealing with incidence properties without referring to magnitudes makes it clear that Karsten was attempting a situational theory of parallels. A full-fledged theory, in fact, was accomplished a few years later in the *Neues System der Parallellinien* (1781) by Carl Hindenburg (1741–1808), which also marked an explicit connection with Leibniz' *analysis situs*. In fact, Hindenburg recalled Kästner's words about the opportunity to employ the notion of *situs* in proving the Parallel Postulate, and claimed that Karsten's conception of *Lage* was nothing other than Leibniz' *situs*. He remained unconvinced, however, by Karsten's proof, and offered a new demonstration of the

a purely positional concept (see above, *note* 93 in § 4). To him, we should at least add Desargues and several other authors working in perspective geometry, who customarily employed a few notions related to the direction of pencils of parallel lines. The direction theory of parallels, in fact, will acquire some relevance in the nineteenth-century foundations of projective geometry, as in von Staudt's *Geometrie der Lage*. It will also have an important role in the birth of vector calculus.

[16] Karsten did not define what *Lage* is, and rather only discussed what does it mean to "have the same *Lage*" (and here the angular considerations are introduced): even though he did not mention Leibniz in this connection, it is very likely that he took this procedure from the similar one employed by Leibniz to define the sameness of *situs* (or *place*) in the correspondence with Clarke (see above *note* 35 in § 3). In the end, Karsten's definition of parallel lines remains similar to Borelli's. As for proving the Parallel Postulate, Karsten's reasoning is the following (see *Theorie von den Parallellinien*, p. 15). Two straight lines have the same *Lage* if they make equal angles with a transversal. Now, since they have the same *Lage*, they will also make equal angles with any other transversal, and … (the Postulate follows). But this is false in hyperbolic geometry, where two straights may make equal angles with a transversal (even two right angles), but unequal angles with another one. We may remark that Borelli and Saccheri had explicitly taken notice of (and avoided) this possible conceptual trap. Karsten answered several criticisms raised against his theory in an essay *Von den Parallellinien und den neuen Bemühungen, die Theorie davon zu ergänzen*, that he included in his *Mathematische Abhandlungen* from 1786 (pp. 113–202). In §§ 19–21 of this work (pp. 132–34), he explicitly compares the *situation* of a straight line with its *direction*, basically claiming that the two notions overlap (but he still distinguishes two opposite directions, toward the right side and toward the left side, in lines having the same situation); in the following pages, Karsten simply writes "*Lage oder Richtung*" as synonyms.

Postulate based on the transitivity of parallelism.[17] Hindenburg's proof of transitivity is wrong, of course, and was attacked several times in the following years; he tried to defend it on a few occasions (especially in a paper from 1786). Hindenburg also declared, however, that if his proof was unconvincing, one might take transitivity as an axiom, since it is by far the most simple assumption that one can make about the theory of parallels. He justified this claim, arguing that the equality of measure is transitive, and thus the identity of *situs* also has to be transitive, since situation and magnitude are together the essential features of any geometrical object. In this respect, Hindenburg was crystal-clear that *situs* should not be conflated with any measure-theoretical notion (including angular width – as in Karsten's theory), and that the concept of parallelism is purely positional.[18]

After Karsten's and Hindenburg's essays, many German mathematicians tried to improve their theory of parallel lines relying on purely non-metrical notions. Among the many papers on the subject, we should mention at least those by Johann Schwab (1743–1821), who authored a booklet that grounded the Parallel Postulate on the notion of *situs* (1801), as well as an entire *Essai sur la situation* (1808) in

[17] Hindenburg's *System* from 1781 was preceded by an essay of criticism of the previous attempts to prove the Postulate, and followed (in the same year) by some *Anmerkungen*. In 1786, he published a new paper, *Noch etwas über die Parallellinien*, in which he again defended his theory; in the same paper, he presented the reader with Lambert's *Theorie der Parallellinien*, which was posthumously edited there. In particular, on Karsten's theory as an *analysis situs*, see *System*, pp. 157–58; on the fact that transitivity may be taken as an axiom, *System*, p. 262 and *Anmerkungen*, p. 348; for Hindenburg's original proof, *System*, pp. 167–68 (it makes no reference to the concept of *Lage*); for the introduction of the notion of *Lage* in Hindenburg's proof, *Anmerkungen*, pp. 347–48; on the concept of *Lage* as a purely positional notion and its reference to Leibniz' *analysis situs*, see *Noch etwas*, p. 383.

[18] We may remark that a proof of the Parallel Postulate through the transitivity of the notion of parallelism was already attempted in the German world by Andreas Segner (1704–1777). Already in his *Vorlesungen über die Rechenkunst und Geometrie* from 1747, Segner had mentioned the notion of *Lage* and *Parallellage* in relation to the theory of parallels (cf. p. 218), even though his development made no use of the new concept nor the notion of transitivity. In his *Cursus mathematicus* from 1756 (which enjoyed a wide circulation in a 1764 German translation as *Anfangsgründe der Arithmetick, Geometrie, und der geometrischen Berechnungen*), however, Segner proved the Parallel Postulate by relying on the *implicit* assumption of the transitivity of parallelism (*Geometria*, § 11; in *Cursus*, p. 144); or, what amounts to the same, that a straight line passing from a point included in an angle must cut this angle (a proposition which is false in hyperbolic geometry, and was later called "Lorenz' Principle", since it was explicitly employed in J.F. Lorenz, *Grundriß der reinen und angewandten Mathematik*, Helmstädt, Fleckeisen 1791). Since the assumption remained implicit, however, Segner did not modify the Euclidean system of definitions, nor did he introduce any concept of direction or situation. The above-mentioned German edition of the same book, however, explicitly took as an axiom the proposition that a smaller angle cannot contain a larger angle (which is also false in hyperbolic geometry), and followed a different way to prove the Parallel Postulate (*Anfangsgründe*, § 259; pp. 198–200). Karsten had accepted Segner's initial proof in his 1758 *Praelectiones* (§ 73; pp. 34–35), and was criticized with him by Klügel (*Recensio*, §§ 11–12, pp. 15–17), even though Klügel's attack on Segner was quite weak (he wrongly criticized Segner's implicit assumption of Aristotle's Principle – which however is not harmful for the proof). Karsten, in any case, was himself unconvinced by his 1758 proof, attempted another solution in 1760 (*Mathesis theoretica*, § 91; pp. 31–35), which was basically the old demonstration by Nasīr ad-Dīn, and finally arrived at his 1778 conception of *Lage*.

which he defended his theory of parallels and tried to renew the old Leibnizian science. Schwab claimed that Wolff's notion of *situs*, that equated it to distance, was a complete misunderstanding of Leibniz' thought (we know that it was, in fact, the only point that Wolff correctly grasped of the entire Leibnizian construction), and that *situs* should rather be taken as a purely non-metrical notion. Schwab, actually, developed an ample theory of *situs*, claiming that it is a simple notion that cannot be defined but only grasped in intuition, since space itself is not a pure concept but a partially sensible representation. Schwab didn't equate *situs* with direction, but explicitly attributed this idea to Karsten. Yet, he defined parallel lines as lines having the same situation and took as an evident principle that two things having the same situation with a third have the same situation between them. The Parallel Postulate is thus once more derived through transitivity. Contrary to Hindenburg's geometrical proof of transitivity, however, Schwab thought that this property (and the Postulate) simply follows *analytically* from the notion of *situs*.[19]

Even though Schwab's books added nothing new to the subject, they were quite important at the time and strongly contributed to the fame of Leibniz' *analysis situs* and his alleged involvement with the theory of parallels. In the same years, the direction theory of parallelism began to acquire significance in the geometrical landscape (merging with the beginnings of projective geometry). This theory, that defined parallels as lines sharing the same direction, was often fashioned after Hindenburg's or Schwab's work, and was thus regarded as a branch of Leibniz' legacy in mathematics. The theory was extremely influential in the nineteenth-century, and dozens of books were published on the topic. A few of them continued to ascribe the invention of this notion to Leibniz' work on *analysis situs*.[20]

---

[19] Schwab's opinion about Wolff's wrong conception of situation is in the *Essai*, § 14 (p. 14), where he also claims that *situs* is not direction (§ 15, p. 15). He had attributed the latter idea to Karsten in his 1801 essay, titled *Tentamen novae parallelarum theoriae notione situs fundatae* (pp. 29–30), where we can also read Schwab's opinions about situation and space (esp. pp. 12–15). The same general remarks are restated in the *Essai*, where Schwab also gives the definition of a straight line through situation in § 22: "Une ligne droite est donc celle dont tous les points ont la même situation" (p. 20). It is very easy to see how naïve this definition is in comparison with Leibniz' own situational definition of a straight line; it is difficult, in fact, to understand Schwab's definition from a geometrical point of view, not to mention employing it in a demonstration. Similarly, parallel lines are defined as straight lines that "ont la même situation entr'elles, ou lorsque la situation de l'une est identique avec celle de l'autre" (§ 28, p. 26), which is a far cry from Leibniz' analogous definition that we have seen, for instance, in *text 26*. The transitivity of parallelism (and thus the Parallel Postulate) is immediately justified (not properly proved) with a comparison with magnitudes: "Si deux grandeurs sont identiques avec une troisième, ells sont identiques entr'elles. Si deux situations sont identiques avec une troisième ells sont identiques entr'elles" (p. 22). In the end, thus, Schwab's philosophical proof of the Postulate is not very different from the one sketched by Proclus about the ὁμοιότης θέσεως (see above, *note* 66 in § 4). The *Tentamen* was not different in this respect, and stated a definition of parallel lines: "Duae lineae rectae in eodem plano jacentes sunt parallelae inter se, si eundem situm habent ad se invicem" (p. 1); and an axiom: "Si duae rectae in plano eundem situm habent ad se invicem; habent etiam eundem situm ad rectam tertiam" (p. 3).

[20] Among the books published in the first half of the century, when Leibniz' influence was still strong (even if immaterial and quite fanciful), see for instance the essay *Die Lehre von den Parallelen* by Carl Vermehren (1816), which discusses at length the notion of *Richtung* (pp. 19–26) and

Following an altogether different line of reasoning, the Kantian mathematician and philosopher Johann Schultz (1739–1805) attempted a proof of the Parallel Postulate in an essay from 1784, which he defended against criticisms in another book on parallels (in 1786) and in a further volume on the infinite (in 1788). Schultz claimed that his mathematical researches on the topic were destined to rediscover Leibniz' *analysis situs* and ground a true geometry of *Lage*.[21] His attempt, however, differs greatly from those we have mentioned so far, and diverges from a direction theory of parallelism. It is grounded, on the contrary, in a new way of employing the actual infinite in geometry. His point of departure is the claim that, since the infinite portion of a plane delimited by any given angle (however small) represents a *finite* part of the whole plane (if you take a right angle, for instance, it is one fourth of the plane), while the infinite portion of a plane delimited by two parallel lines (a strip, however large) represents an infinitesimal part of it (since you need an infinite number of such strips to cover the entire plane), then any angle whatsoever is *bigger* than any strip whatsoever (despite the fact that they are both of infinite measure). Granted this, Schultz easily proves the Parallel Postulate: he considers two straight lines *a* and *b* meeting with a transversal line *c* at a right angle α and an acute angle β (in P), and draws the parallel *d* to *a* in P. If *b* did not meet *a*, then the angle γ formed by *b* and *d* would be entirely included in the strip bounded by *a* and *d*, and this is impossible since the angle is bigger than the strip; thus *a* and *b* meet, proving the Postulate.

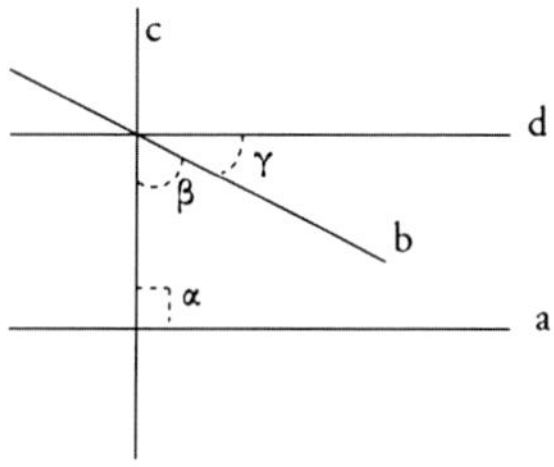

It is highly remarkable that Leibniz *had* tried to prove the Parallel Postulate in a somewhat similar way (see especially *texts 3* and *22*), employing actually infinite strips of the plane, indulging in comparing infinite measures among them, and trying to apply to infinite extensions the axiom that the whole is bigger than the part. Schultz could not possibly have been influenced by

assumes an axiom stating the transitivity of the sameness of direction (p. 21); or Andreas Jacobi's book on Euclid's Eleventh Axiom (1824), which at §§ 53–54 (pp. 36–41) also explicitly deals with the concept of direction; or the essays by Joseph Knar (1827–1828), that tried to define the notion of an angle through the concept of direction (and then prove the Parallel Postulate, of course). The geometrical textbooks dealing with a direction theory of parallelism in the second half of the century are a legion. For a few titles and a discussion on the matter, see SCHOTTEN 1893 and VOLKERT 2013. The definition of parallels through direction was famously criticized in the book of Lewis Carroll on *Euclid and His Modern Rivals* (DODGSON 1879, pp. 70–131), and shortly thereafter (1884) by Frege in § 64 of *Grundlagen der Arithmetik*. A thorough examination of Frege's views and sources on the topic is to be found in the recent paper MANCOSU 2014. We have already mentioned (*note* 5 in § 1) that one of the few references to Leibniz' writings on the Parallel Postulate is to be found in Wilhelm Killing's *Grundlagen der Geometrie*, where Leibniz' theory of parallelism is (very mistakenly) identified with a theory of direction.

[21] J.F. SCHULTZ, *Entdeckte Theorie der Parallelen* (1784); *Darstellung der vollkommenen Evidenz und Schärfe seiner Theorie der Parallelen* (1786); *Versuch einer genauen Theorie des Unendlichen* (1788). Schultz' reference to Leibniz' *analysis situs* (through the well-known Kästner's quotation in Klügel's *Recensio*) is in the *Vorrede* to his *Entdeckte Theorie der Parallelen* (esp. pp. 18 and 30–31).

these Leibnizian papers, though, since none of them was published at the time. His reference to the *analysis situs* in this connection has to be regarded as purely incidental; and Leibniz himself had probably never connected these infinitistic attempts with the notion of situation. This demonstrative strategy, if the truth be told, was natural enough in proving the Parallel Postulate that the French mathematician Louis Bertrand had independently devised a similar proof in the same years as Schultz.[22] It is highly remarkable, in any case, that Schultz was among the first German mathematicians to explicitly state that geometry should be a science of space (a *Raumlehre*).[23]

Schultz' proof, quite understandably, was attacked by many mathematicians for his liberal use of the actual infinite, and Schultz built an entire theory on the topic hoping to save his own demonstration of the Postulate; he even devised a few other proofs based on the same principles. Men like Kästner and Hindenburg strongly opposed Schultz' theory, while several other German mathematicians endorsed his views.[24] Since Schultz was a renowned Kantian philosopher, moreover, from time to time his theses were associated with Kant's epistemology of mathematics, and earned followers for this very reason; meanwhile Kant himself, embarrassed by his friend's theory, tried to disentangle his philosophy from Schultz' views.[25] Schultz'

---

[22] Bertrand's attempt is to be found in his *Developpement nouveau de la partie elementaire des mathématiques* (1778), vol. 2, pp. 19–24. Among the French mathematicians who followed him, we may recall Gergonne and Legendre; the latter accepted and improved Bertrand's proof in his *Réflexions sur différentes manières de démontrer la théorie des parallèles*, pp. 397–407.

[23] Schultz' ground for this claim was the possibility of having a geometrical theory of the infinite, and thus of space itself as an infinite whole. This idea put him in contrast with several mathematicians, and with his philosophical master himself, Kant (see the *notes* below). A few statements about Schultz' new *Raumlehre* are to be found in his *Darstellung seiner Theorie der Parallelen*, pp. 46–47; and in *Versuch einer genauen Theorie des Unendlichen*, pp. 14 and 44. On the other hand, the champion of the Leibnizian party in the last years of the eighteenth century, Johann August Eberhard (1739–1809), explicitly denied that Leibniz' new conception of space as a situational order might ever have any relevant consequence in the methods of geometry (!): see his essay *Ist die Mathematik* ..., where he claimed: "Die Meinung von Leibniz, er definirte den Raum durch die Ordnung der zugleich- und auseinanderseyende Dinge; … In der Geometrie ist dieser Begriff des Raumes freilich unbrauchbar" (in the 1790 editin of the *Philosophisches Magazin*, vol. 3.1, p. 99).

[24] Kästner's and Hindenburg's criticisms of Schultz may be read in Hindenburg's *Noch etwas*, pp. 392–401. Kästner, in fact, was opposing Bertrand's essay. In any case, many more mathematicians discussed and rejected Schultz' method, for many different reasons; and since it presented something really new to the mathematical community, the discussion was carried on well into the nineteenth century. Among the German followers of Schultz, we should mention at least Crelle (in 1816) and Hessling (in 1818), whose infinitary proofs, however, have less points of contact with Leibniz than the original one by Schultz.

[25] Schultz published two very important works of interpretation and defense of Kantian philosophy, the *Erläuterungen über des Herrn Professor Kant Critik der reinen Vernunft* (1784) and the *Prüfung der Kantischen Critik der reinen Vernunft* (1789–1792). Moreover, the aforementioned (see *notes* 1 in § 4, and 12 here) articles by Kästner on the nature of space were published in Eberhard's pro-Leibnizian and anti-Kantian *Philosophisches Magazin*, and were intended by Eberhard as a side-attack on Kant's philosophy. Given the great authority of Kästner in Germany, Kant decided to answer the criticisms and gave Schultz a few of his notes to be developed into a polished paper. Schultz reworked the Kantian manuscript and published a formal reply to Kästner (which is now to be found among Kant's works: *Über Kästners Abhandlungen*, KGS X, pp. 410–23). Since Kästner

theory on measuring an actually infinite space may have sounded truly Leibnizian to many mathematicians of the time, who identified Leibniz with the discoverer of the Calculus, but could not have known anything about Leibniz' real endeavors in this direction. These infinitary methods were eventually completely superseded by the progressive advancement of a "finitistic" conception of differential analysis that characterized the nineteenth century.[26]

5.4 The Principle of Reason.

These German essays on situation and parallels were often intermingled with important epistemological considerations about Leibniz' overall geometrical project. The main issue was probably that of logical analysis, and whether mathematical propositions may be proved through a simple analysis of the concepts involved (as Leibniz claimed) or whether they needed some appeal to intuition (pure or empirical) or other forms of reasoning. Schultz accused Karsten's and Hindenburg's theories of parallels of being unsatisfactory since they only relied on a "philosophical" analysis of the notion of *Lage*, while his own proof was truly "mathematical" and grounded on synthetic *a priori* judgements.[27] On this point, Kant himself was probably less convinced than his friend.[28] Hindenburg strongly rebutted the allegations, and claimed that his proof, and that of Karsten as well, were properly mathematical: a philosophical proof was rather the one given by Schultz, that relied on an argument employing the infinite. Karsten himself, on the other hand, endorsed a vaguely Wolffian epistemology, and probably considered his own proof as truly

had attacked his theory of parallels, however, Schultz added a conclusion to the paper, defending his conception of the actual infinite, which was very different from Kant's more traditional opinion about it. Kant strenuously protested with Schultz for this treatment, which approached too much his own Criticism and Schultz' un-Kantian and highly objectionable theory of the infinite (see Kant's letters to Schultz from August 2nd and 26th, 1790; KGS XI, pp. 184 and 200–201).

[26] We may note that Schultz' theory of the infinite was later discussed by Bolzano in his *Paradoxien des Unendlichen* (1851), § 44, pp. 85–87. A modern reconstruction of Schultz' doctrine is to be found in SCHUBRING 1982.

[27] Schultz' opinion against Karsten and Hindenburg was stated in his *Entdeckte Theorie der Parallelen*: "Dass es aber überhaupt schlechterdings unmöglich sey, die Lehre von den Parallellinien durch blosse Analysirung des Begrifs ihrer Lage festzustellen, …" (p. 41).

[28] The philosophical character of the proof of the Parallel Postulate grounded on the notion of *Lage* is the main topic of Kant's own *Reflexionen* on the matter. See for instance Kant's *Reflexion* 9, that concludes: "Vielleicht ist dies ein Satz für die Geometrie der Lagen" (KGS X, p. 37). We don't know exactly which essays on the Parallel Postulate Kant had read when he was working on the topic; he was probably drawing on Schultz' books. In the end, however, Kant accepted a "philosophical" proof of the Postulate that could dispense with employing the actual infinite ("ohne herbeyziehung unendlicher Flächen", KGS X, p. 45). It is highly peculiar that while the hyper-logicist Leibniz may have accepted (in his private writings) a non-analytical demonstration of the Parallel Postulate, grounded on the Principle of Reason, Kant may have accepted (in his very private writings) a non-synthetic proof of it, relying on the simple analysis of the concept of *Lage*. This frankly preposterous reversal of their common epistemological positions shows once again how important the story of the attempts to prove the Parallel Postulate has been for the history of philosophy (and vice versa).

analytical (and Hindenburg's unrequested defense quite misplaced).[29] Schwab, in any case, who had devised his own very philosophical proof of the Postulate, entirely grounded on conceptual analysis, launched a wide-ranging attack against synthetic *a priori* propositions and the use of intuition in mathematics. He was in turn criticized by Gauss himself, who in his only public declaration about the Parallel Postulate stated that Schwab's analysis of the concept of *situs* was a mathematical aberration.[30] Among these epistemological discussions, which prepared and nurtured the birth of non-Euclidean geometry, we may also mention that Karsten saw in Hindenburg's attempt to ground the Parallel Postulate directly on the transitivity of parallelism a kind of recourse to the Principle of Reason, here invoked (in Karsten's eyes) to guarantee that transitivity has *no reason* to be violated. Hindenburg replied by denying the charge, since the Principle of Reason (as Leibniz himself stated) cannot have any use in mathematics, and one cannot hope to prove the Parallel Postulate by relying on this principle.[31]

This standard opinion about the restricted scientific fields in which the Principle of Reason could be legitimately and successfully employed (which did not include the purely mathematical sciences) prevented almost any attempt of a proof of the Parallel Postulate according to this principle. An exception in Germany is represented by the mathematician Martin Ohm (1792–1872), who in 1819 presented a proof of the Postulate that was basically the same as Schwab's, but explicitly mentioned the Principle of Reason. Since two parallel lines have the same *Lage*, Ohm said, and the latter is defined quite independently of any transversal line cutting them, then there is *no reason* why such a transversal line that meets both parallels should make different angles

[29] See for instance Hindenburg's defense against Schultz' criticisms in *Noch etwas* pp. 392–97. In the *Anmerkungen*, Hindenburg claimed that his own proof "sich nicht aus Analysierung des Begriffs von Lage gründet" (p. 348). Hindenburg's criticisms to Schultz' "philosophical" proof are to be found in *Noch etwas*, p. 368.

[30] Schwab's main criticisms of Kant's conception of intuition and synthetic *a priori* judgments, and his unyielding assessment of geometry as a purely analytical discipline, are mostly voiced in his late essay *Commentatio in primum elementorum Euclidis librum* from 1814. As we have seen in a few previous quotations, Schwab was not a pure "Platonist", though; he was almost an Aristotelian in this respect, and believed that the original concept of *situs* was first given in (empirical) intuition. Gauss reviewed Schwab's *Commentatio* in a 1816 issue of the *Göttingische gelehrte Anzeigen* (now to be read in Gauss' *Werke*, vol. 4, pp. 364–68), complaining about the emptiness of the notion of *Lage* and the purely philosophical proof attempted by Schwab. Similar criticisms of Schwab's analytical proof of the Parallel Postulate (as it was presented in the *Tentamen*) were already advanced in Schweikart's *Theorie der Parallellinien* from 1807 (pp. 25–26), which directly referred to Kant's epistemology of synthetic judgements to show the inanity of Schwab's way of proof. Gauss' reference to intuition in the above-mentioned review was sometimes read (possibly thanks to Schweikart's opinion) as an endorsement of Kant's conception of synthetic *a priori* judgments; even though this was probably not the original intent of Gauss (who was more an empiricist than a follower of Kant), Gauss' statement was later considered by neo-Kantian philosophers as an authoritative opinion about the possibility of reconciling Critical philosophy with non-Euclidean geometry. Gauss' few papers on non-Euclidean geometry are collected in Engel&Stäckel 1895 and commented in Reichardt 1985.

[31] See Hindenburg's *Noch etwas*, pp. 373–74.

with the two.[32] A similar stance was taken by the astronomer and mathematician Jakob Lehmann (1800–1863), who in his *Mathematische Abhandlungen* attempted a new justification of the Postulate based on Wallis' principle of similarity, complementing and improving (he believed) the old proof by arguing that there is no sufficient reason why transformations by similarity should be impossible in space; therefore the Parallel Postulate holds.[33] Lehmann added that the employment of the Principle of Reason in mathematics should not be the rule, but an exception in case of need; and perhaps in the future we would be able to dispense with it. He never doubted the truth of the Postulate, even though he speculated a bit on the consequences of its denial.[34] As we easily see, both Ohm's and Lehmann's proofs are a simple rephrasing of older demonstrations with an epistemological twist invoking the lack of a sufficient reason.

The epistemology behind the Principle of Reason, moreover, prevented for several years a promising connection between the foundations of geometry and those of statics. The very first propositions in statics depend, in fact, on the theory of parallels. More specifically, the Postulate is needed to prove that the resultant of two equal forces having the same direction, applied at right angles to the extremities of a segment, is equivalent to a force of double intensity applied to the center of the segment.[35] A different principle grounds, in fact, a consistent and rich non-Euclidean science of statics. Thus, for instance, if we take an ideal, equal-armed, weightless balance, and we suspend a weight of 2 to one end of it, while on the other end we suspend at a right angle another ideal, equal-armed balance with two weights of 1 at each of its ends, the entire machine will be in equilibrium if and only if the Parallel Postulate holds. Since Leibniz himself had used the Principle of Reason to justify the Archimedean axiom on the equilibrium of the simple balance, it would be quite natural to try to employ it to the same effect on this composite balance. This, on the other hand, would represent a proof of the Parallel Postulate by sufficient reason.[36]

---

[32] The essay by Ohm is the *Kritische Beleuchtungen der Mathematik*, which has a discussion on parallel lines at §§ 43–44 (pp. 31–34). The aforementioned reasoning is introduced by: "II. Schneiden sich zwey Linien nie, werden sie aber durch eine beliebige dritte Linie geschnitten, so sind die äußern Winkel den innern gleich. … Dieser zweyte Satz folgt unmittelbar aus dem Satz des zureichenden Grundes …" (p. 32).

[33] Lehmann, *Mathematische Abhandlungen betreffend die Begründung und Bearbeitung verschiedener mathematischer Theorieen* (1829), pp. 335–41.

[34] Here the quote about the Principle of Reason: "Der Satz des zureichenden Gründes soll für den Nothfall verspart; also so selten als möglich angewandt werden, damit die eigenthümliche, intuitive, geometrische Methode, ohne hülfe dieses Gründsatzes, soviel als möglich wirke" (*Mathematische Abhandlungen*, p. 337). Lehmann knew several attempts to prove the Parallel Postulate that we now range among the prehistory of non-Euclidean treatises, such as Lambert's and Taurinus'. He thus speculates a bit about hyperbolic geometry as an *imaginary* geometry (in the proper sense). On Lehmann's foundations of geometry, see Pont 1986.

[35] I follow here the discussion in Bonola 1906, pp. 173–76 (English ed., pp. 181–84).

[36] To my knowledge, the first person to conceive the example of the composite balance and examine its conditions of equilibrium was Christiaan Huygens in a paper from 1693 titled *Demonstration de l'equilibre de la bilance* (now, along with its unpublished drafts, in Huygens' *Oeuvres completes*, vol. 19, pp. 37–47). Huygens was not interested in proving the Parallel Postulate, but in grounding the much discussed Proposition 6 of the First Book of Archimedes' *De aequilibriis planorum*

No one, however, realized this connection for a long time, and in any case the use of sufficient reason in pure geometry remained excluded because of epistemological considerations.[37] The Leibnizian physicist Samuel König, for instance, grounded his work on statics on the Principle of Reason, but could not conceive of violating the orthodox Leibnizian view that geometry should only make use of the Principle of Contradiction. He was thus at pains to disentangle the specific contributions of the two principles in his work; with dubious results, and in any case missing the opportunity to explain their connection.[38] The first to realize the link between the Parallel Postulate and the laws of statics was probably Joseph Fourier (1768–1830), who in an unpublished manuscript from the 1820s tried to prove the Postulate, and ended up grounding it on the above-mentioned principle on the composition of forces. He claimed, in fact, that the Ancients had failed to solve two problems, the foundation of statics and the theory of parallels, which he proved to be reducible to one question. He had no Leibnizian biases, however, and concluded his work with some reflections moving toward an empiricist conception of geometry, in which the truth of the Parallel Postulate was founded on *a posteriori* grounds.[39]

Among the many who missed the relation between statics and the theory of parallels, we may also count Joseph-Louis Lagrange (1736–1813), who in his work on mechanics dealt with the equilibrium of the aforementioned composite balance without invoking any principle on parallel lines.[40] Lagrange, however, also devised an

(Archimedes, *Opera*, ed. Heiberg, vol. 2, pp. 132–36). Huygens does not stress that the Euclidean Postulate is the pivotal mathematical assumption of his demonstration; in particular, in the proof of Prop. 3 (pp. 45–47), he essentially employs similar triangles – that do not exist in hyperbolic space. It is very likely that Leibniz knew Huygens' short essay. Moreover, it was first written and presented to the French Academy of Sciences in 1672 and 1673, when Leibniz was studying in Paris as a pupil of Huygens. I do not know, however, of any use of this result of Huygens in Leibniz' mathematical writings.

[37] We may mention that Saccheri himself, who devoted his entire life to a proof of the Parallel Postulate and authored a book on statics (the *Neo-Statica*, see above, *note* 15 in § 2) completely missed any connection between his two major mathematical efforts. Lambert, too, seems to have overlooked their link; on this topic, see especially Gray&Tilling 1978.

[38] See especially König's infamous paper *De universali principio aequilibrii* from 1751, which started the quarrel with Maupertuis that would eventually bring down König's own reputation (see above, *note* 1). Here König is visibly embarrassed while trying to justify a principle of statics which is partly geometrical and partly physical; he employs the Principle of Contradiction to prove that several forces together may result in a static equilibrium (something that Kant will attack in 1763, claiming that logical analysis based on the Principle of Contradiction is not strong enough for this), and the Principle of Reason to prove that a static equilibrium cannot be broken unless a new force is applied (p. 167); cf. however a similar stance in Leibniz' letter to Mariotte quoted in *note* 57 in § 3). König also wrote an edition of the first six books of the *Elements*, that was posthumously published in 1762, in which he straightforwardly assumed a form of the Parallel Postulate as an axiom.

[39] Fourier's studies on the Parallel Postulate are still unpublished, but are widely discussed in Pont 1986, pp. 533–86. His empiricist outcomes are clearly stated: "Ainsi la géométrie reposerait aussi comme la statique sur une proposition de fait connue de tous les hommes par une expérience constante. … C'est pour cela que la conaissance expérimentale que tous les homes acquirèrent des proprietés de l'espace sert de fondement aux théorèmes de la géométrie" (Pont 1986, p. 582).

[40] In his celebrated *Mécanique analytique*, Lagrange gives an explanation of the Archimedean principle of statics with a reference to the absence of reasons why a balance should incline on one

independent proof of the Parallel Postulate entirely grounded on the Principle of Reason, which he presented to the Academy of Sciences in 1806. In it, he proved the uniqueness of the parallel line through a point (Playfair's Axiom) by means of a symmetry principle. Let *b* be the parallel line to *a* through P; should a second straight line *c* be parallel to *a* through P, forming with *b* a given angle β, then since by the definition of parallel lines "everything must be equal on their two sides" and there is *no reason* why we should find any difference in their situations, it would be possible to take one more straight line *d*, which has the same situation with *c*, as *c* with *b*. This line *d* would form an angle γ with *c*, equal to β, and would be itself parallel to *a*. Repeating the reasoning, we easily arrive at a contradiction. In fact, if we take a point Q on *a* and join it with P we see that any straight line *e* falling in the triangle OQP cuts the line *a*, and thus there is an angle at P under which the straight lines *a* and *e* meet; but (with some care) we can always reflect our "parallel lines" *b*, *c*, *d*, etc. as to have one of them falling into the triangle.[41]

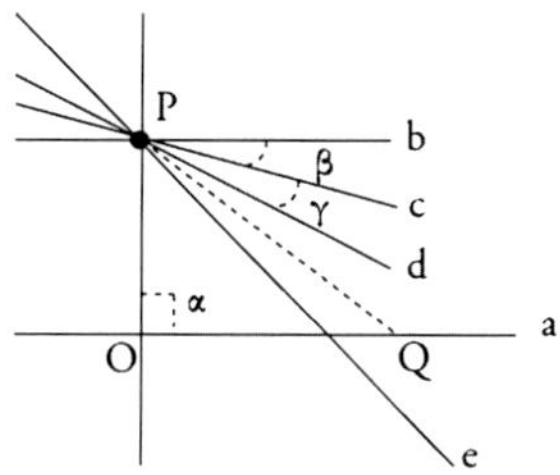

The similarity of this proof with Leibniz' own attempts in *In Euclidis* πρῶτα is quite striking. Lagrange's reference to a definition of parallel lines (which he never spells out), given through their identical behavior on their two sides, closely recalls Leibniz' statement about lines that "*ubique eodem modo se invicem habent*", which we have seen at work in

side in case the arms and the weights are equal; he does not here explicitly mention, however, the Principle of Reason (see *Oeuvres*, vol. 11, p. 4). He goes on to discuss the case of the composite balance that we have mentioned earlier (the one examined by Bonola), but again he refers neither to the Parallel Postulate nor to the Principle of Reason to justify the condition of equilibrium. In fact, Lagrange's discussion of the topic has a markedly Greek and Archimedean flavor, as he considers the equilibrium of a triangle rather than that of a composite balance (they amount to the same thing); in doing this, he implicitly employs a consequence of *Elements* VI, 2 (Thales' Theorem) that Leibniz (and Saccheri) had discussed in relation to the theory of parallels; see above, *text 2*, and *note* 21. In any case, he recalls Huygens' studies on the matter. On Lagrange as a historian of mechanics, see Oliveira 2013.

[41] Lagrange's proof can be read in Borgato&Pepe 1988. The main passage is the following, in which Lagrange states that the Principle of Contradiction is strong enough to prove all geometrical axioms (he gives the example of the whole bigger than the part!) except the Parallel Postulate, which needs another principle, that of Sufficient Reason, which is the foundation of statics: "Par un point donné on ne peut tirer qu'une seule droite parallele à une droite donnée en position. Voilà le principe sur le quel repose la theorie des paralleles ; mais peut on prendre ce principe pour un axiome evident de soi meme ? Nous observerons que les axiomes recus en geometrie sont des propositions qui peuvent se demontrer par le seul principe de contradiction ; ainsi le tout est plus grand que sa partie parce qu'il y a contradiction a ce que la partie egale le tout ; et on peut dire que toutes les propositions de geometrie qui ne dependent point de la théorie des paralleles peuvent etre demontrées par le seul principe de contradiction. Mais il y a un autre principe egalement evident par lui meme ; c'est celui de la raison suffisante auquel Leibniz avait peut etre donné trop d'extension, mais qui reduit a sa juste valeur fournit aux demonstrations mathematiques une base aussi solide que celui de contradiction. C'est sur ce principe qu'Archimede a etabli la theorie de l'equilibre dans le levier ; et que toute la statique est fondée" (pp. 331–32). On Lagrange's proof, see also Grabiner 2009. We may remark that it employs Pasch's Axiom in stating that line *e*, entering the triangle at P, will cut it again at a point in *a*; on Leibniz' and Pasch's Axiom, see above, *note* 13 in § 4.

his papers from the 1690s and 1700s. Lagrange's employment of the Principle of Reason as a symmetry in space, too, is almost identical to Leibniz' use of it. In particular, Lagrange's line of reasoning is the same as we find in Leibniz' attempt to prove the axiom that two straight lines do not enclose space, which is the only other occurrence, besides the Parallel Postulate, in which Leibniz resorted to the Principle of Reason in a geometrical theorem (again in *In Euclidis* πρῶτα). Lagrange himself remarked that a similar proof may be easily devised to prove that axiom.[42]

It is hard to speculate about an actual acquaintance of Lagrange with Leibniz' unpublished papers. We know that Leibniz' editor Louis Dutens, who published the most important collection of Leibniz' works in the eighteenth century (six volumes in 1768), asked Lagrange to edit Leibniz' papers on mathematics. We don't know whether Lagrange ever began this work or had any actual contact with Leibniz' mathematical manuscripts. We do know that he resigned from this duty, and Dutens, lacking an expert editor, simply republished those few mathematical papers by Leibniz that he had published during his lifetime.[43] It remains a hard guess to suppose that Lagrange may have seen the manuscript of *In Euclidis* πρῶτα in the 1760s and forty years later he presented a similar paper on the Parallel Postulate. Nevertheless, it is remarkable how a mathematician deeply involved in Leibniz' mathematical constructions and metaphysics (as Lagrange surely was), may have re-invented a proof of the Parallel Postulate based on the very same principles and the same geometrical development.

In any case, Lagrange seems to have been unsatisfied by his attempt, and the Academy's cold reception of his paper shows that the times were already changing. In the late 1820s and the early 1830s the works of Lobachevsky and Bolyai dismissed the attempts to prove the Parallel Postulate and began the most rewarding construction of hyperbolic geometry. Their discoveries and projects remained at first unknown or encountered strong opposition, and we may count several attempts to prove the Postulate in the second half of the nineteenth century. The publication of Gauss' cor-

[42] Here the quote by Lagrange: "Nous observons ici que ce n'est que par le meme principe, qu'on peut demontrer la proposition que les Geometres modernes prennent pour axiome, et qui consiste en ce qu'on ne peut mener par deux points donnés qu'une seule ligne droite. Car si une droite etant menée par deux points on pourroit en mener une autre, elle seroit necessairement d'un coté de la première ; donc on pourroit encore en mener une troisième de l'autre coté, et ainsi de suite. Par la on prouve aussi l'axiome d'Euclide que deux droites ne peuvent renfermer d'espace, laquel ne paroit pas pouvoir se deduire du principe de contradiction" (Borgato&Pepe 1988, p. 333).

[43] Lagrange informed D'Alembert about his involvement in the Leibniz edition in his letter to him from January 26th, 1765, which may be read in the edition of Lagrange's correspondence in his *Oeuvres*, vol. 13, pp. 31–32; in the letter, Lagrange already doubts that he will be able to comply with Dutens' requests. D'Alembert, however, answers Lagrange by sending him an original letter written by Leibniz (to Varignon), which he happened to possess. Lagrange thanks his friend for the Leibnizian manuscript saying that he will place it "dans le recueil des pieces mathématiques de l'auteur, que j'ai entre les mains" (March 20th, 1765; *Oeuvres*, vol. 13, pp. 36–37). We have no further information about this "recueil", which probably contained several manuscripts (as Leibniz' correspondence, at least, was in manuscript form). Dutens informs the reader about Lagrange's impossibility of carrying on the task of an edition of Leibniz' mathematical writings in his Preface to the 1768 Leibniz edition (vol. 3, p. 1). On Dutens' edition of Leibniz, see Heinekamp 1986. More in general on Lagrange's attitude toward Leibniz, see Taton 1986.

respondence, and then Riemann's, Helmholtz', Klein's works were needed (among others) to foster the acceptance and development of non-Euclidean geometry in the 1860s and 1870s. We won't follow, however, this further story. The late-coming attempts to prove the Postulate had no real impact on the history of mathematics, nor do they usually refer to Leibniz, whose metaphysics and epistemology had ceased, in the new century, to be active intellectual forces. The full-fledged development of non-Euclidean geometry and the philosophical reflections on it (mostly empiricist or neo-Kantian) were even further removed from Leibniz' mathematical and epistemological views. When Leibniz' logical papers were rediscovered at the beginning of the twentieth century and a new neo-Leibnizian philosophical season began, the question of parallel lines had been already settled and the epistemological disputes about non-Euclidean geometries mostly concerned the choice between empiricism and conventionalism, or the interpretation of the General Theory of Relativity. It is thus not entirely surprising that Leibniz' papers on the Parallel Postulate attracted no interest from the community of scholars. Nevertheless, they represent an important episode in the history of thought.

II Leibniz' Texts on Parallel Lines

6

6.1 A vii, 1, n. 9, p. 109.

Parallelae sunt lineae aequidistantes.

6.2 A vii, 1, n. 34, p. 208.

Si a tribus eiusdem rectae punctis, quorum extrema aeque a medio absunt tres alteri rectae occurrant parallelae, summa extremarum duplae mediae aequalis erit. In recta AC medium punctum B unde ad rectam DE, parallelae, AD, BE, CF, aio esse AD + CF aequal. 2 BE. Per D punctum quo aliud F propius non sit transeat DLM parallela ipsi ABC, occurrens rectis BE, CF productis si opus, in punctis L, M. Patet aequales esse AD, BL, CM, adeoque duplum mediae aequari summae extremarum …

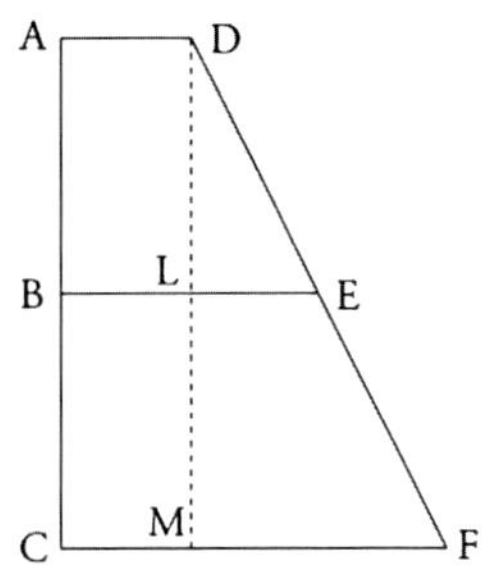

V. De Risi, *Leibniz on the Parallel Postulate and the Foundations of Geometry*,
Science Networks. Historical Studies, DOI 10.1007/978-3-319-19863-7_6

Leibniz' Texts on Parallel Lines 6

6.1 From *Characteristica geometrica. Reading Notes to Fabri's "Synopsis geometrica"* (1673)

Parallels are equidistant lines.

6.2 From *De secandis parallelis* (1676).

If, from three points on the same straight line, of which the extremes stand apart from the middle equally, three parallels meet another straight line, the sum of the extreme lines will be equal to twice the middle line.

Let B be the midpoint on line AC, from which the parallels AD, BE, CF meet the straight line DE, I say that (AD)+(CF)=2(BE).

Let DLM, parallel to ABC itself, pass through point D, which is not the closest to F, meeting the straight lines BE, CF (produced if necessary) in points L, M. It is obvious that AD, BL and CM are equal, and moreover that the double of the middle line equals the sum of the extreme lines …

[*The text goes on with the demonstration, and adds a few elementary results. The diargam has been simplified*]

V. De Risi, *Leibniz on the Parallel Postulate and the Foundations of Geometry*, Science Networks. Historical Studies, DOI 10.1007/978-3-319-19863-7_6

6.3 A VI, 3, nn. 59, 65, 66, pp. 471, 487–89, 489–90.

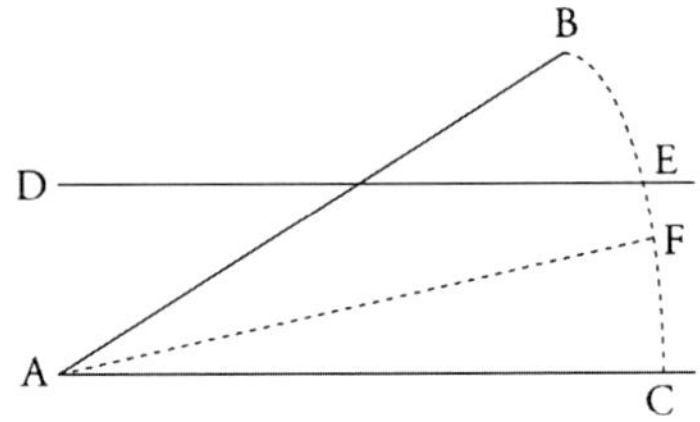

Linea infinita est immobilis. Sit linea AB infinita a parte B quae motu transferenda sit in AC. Sit inter B et C ipsa DE parallela AC. Quando perveniet in AC, erit tota infra DE, et in quocunque puncto ponatur, ut in AF, erit pars ejus infinita supra DE. Unde si AC ponatur perfecte interminata, seu si nullum sit punctum ultimum, necesse est, ut tandem simul tota illa linea interminata infra DE descendat, totaque spatium interjectum simul conficiat, id est ut sit in pluribus locis. Ἄτοπον.

...

Videndum an necesse sit omnium Parallelarum ejusdem plani media puncta esse in una recta (eaque ipsis normali). Hoc posito omnium datae parallelarum in aliis planis omnibus sumtarum puncta media cadent in planum ipsis normale. Unde sequitur planum aliquod dari quod bisecet universum. ... Sed probandum adhuc restat planum inter duas parallelas interminatas interceptum, interminatae longitudinis sed terminatae latitudinis, esse bisecabile per aliquam rectam extremis parallelam. Equidem ubi maius et minus et omnia utrinque paria, videtur et aequale, sed sine assumto motu, probatio minus manifesta. Ex quo motu illud sequitur in plano cuilibet rectae aliam duci posse parallelam planum bisecantem, eamque non nisi unam. Hinc sequitur duas rectas planum bisecantes se secare. Haec de terminatis et interminatis vera, et in terminatis de trisecantibus. Interminata quantitas in plures duabus partes aequales secari non potest.

...

Extensio interminata non debet implicare, quia videntur aliqua de ea demonstrari posse, ut duas rectas interminatas in eodem plano quae non sint parallelae, unum habere punctum commune. Quod de terminatis dici non potest. Sed hoc tamen de terminatis dici potest, produci posse dum concurrant. Videtur vero intelligi recta jam producta; imo rectae per se interminatae a nobis aut corporibus terminantur.

6.3 From *Linea infinita & extensio interminata* (January–April 1676)

Translated in Arthur 2001, pp. 41, 53, 69–71.

[A] An infinite line is immovable. Let there be a line AB, infinite in the direction B, which is to be translated by a motion to AC. Between B and C let there be a line DE parallel to AC. When it reaches AC, it will be completely below DE, and at whatever [*intermediate*] point is assumed, for example at AF, there will be an infinite part of it above DE. Whence if AC is assumed to be perfectly unbounded, i. e. if there is no last point, it is necessary that the whole of that unbounded line should finally descend below DE at the same time, and the whole of it would complete the intervening space simultaneously, i. e. be in several places at the same time – an absurdity.

…

[B] We must see whether it is necessary for the midpoints of all the parallels in the same plane to be in one line (and for this to be normal to them). Supposing this is so, the midpoints of all the lines parallel to the given one assumed in all the different planes fall in a plane normal to them. Whence it follows that there is some plane that bisects the universe. … But it still remains for us to prove that the plane intercepted between two parallel unbounded lines, having an unbounded length and a bounded breadth, is bisectible by some straight line parallel to the outermost ones. For my part it seems that, whenever greater and less and everything on each side are alike, they are also equal, but without the assumption of motion the proof is less evident. With the assumption of motion, it follows that to any straight line in a plane another can be drawn parallel to it bisecting the plane, and only one such parallel may be drawn. Hence it follows that two straight lines bisecting the same plane intersect. These things are true about both bounded and unbounded things, and in bounded ones they are also true about lines trisecting the plane. An unbounded quantity cannot be cut into more than two equal parts.

…

[C] An *unbounded extension* ought not to entail a contradiction, since it seems that some things can be demonstrated about it, for instance, that two unbounded straight lines in the same plane that are not parallel have one point in common. And this cannot be said of bounded ones. What can be said about bounded lines, however, is that it is possible for them to be further produced until they meet. Indeed, it seems, a straight line is understood as being already produced; or rather, straight lines, unbounded in themselves, are bounded by us or by bodies.

6.4 A VII, 6, n. 14, 20, pp. 144, 211.

Rectissime enim a Pascalio in Conicis admonitum est convergentes rectas non minus censeri debere ordinatas quam rectas parallelas. Cum et parallelae possint species convergentium haberi, quarum punctum concursus infinite absit.

...

Sed hoc non aliter fieri potest (ne quis hic erret) nisi ponatur recta μλ non esse ipsi AT omnino parallela, sed ad eam nonnihil inclinata, nec proinde curvae Dδ asymptotos, sed ei occurrens, alicubi ut in λ, licet λ absit infinito abhinc intervallo. Id est recta μλ erit quidem *infinita* sive quavis designabili major, sed non interminata.

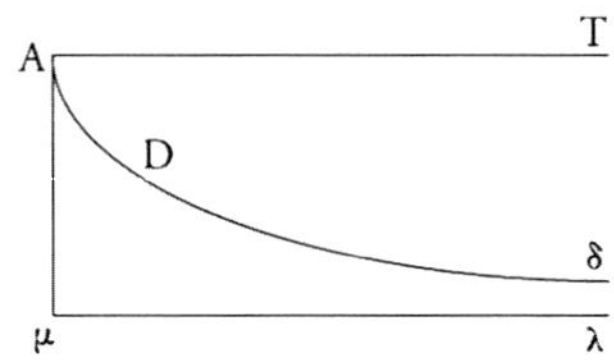

6.5 Niedersachsische Landesbibliothek, Leibn. Marg. 184; pp. 5, 8.

In genere *parallelae* quarum eadem ubique distantia est.
Hoc Axioma est difficile. Imò nec verum est nisi rectae sint in eodem plano.

6.6 Echeverría 1995, pp. 58–60.

Axioma eius 11 est: si in duas rectas CDF, ABG incidat tertia EDB, sitque angulus FDB+GBD minor duobus rectis, erit H punctum concursus ad partes F, G. Hoc pro axiomate sumtum miror, quis enim non videt quam parum sit notum, et quam multa difficiliora supponat.

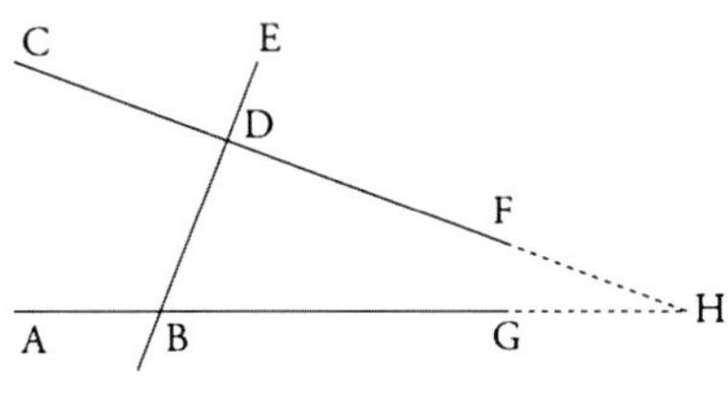

6.7 Echeverría 1995, p. 91.

Duae rectae concurrentes productae sese secabunt. Hoc demonstrari non potest nisi ipsius consideratio plani praecesserit. Considerandum enim duas rectas punctum commune habentes esse in eodem plano rectaque in plano indefinita existente partem plani ab hac alteram ab altera rectae parte esse. Et rectam quae ad alteram accedit continuo ad alteram partem accedere uniformiter adeoque necessario ad eam pervenire.

6.4 From *Dissertationis de arithmetica circuli quadratura propositiones septem and Quadraturae circuli arithmetica pars prima* (April–June 1676)

[A] It was admonished most correctly by Pascal in the *Treatise on Conics*, that convergent straight lines must no less be judged to be ordinates than parallel straight lines, since parallel straight lines can be considered a kind of convergent straight lines – those of which the point of intersection stands apart infinitely.

…

[B] But this cannot be otherwise (so that no one may be here mistaken) unless the straight line μλ is posited not to be perfectly parallel to AT, but inclined to it somewhat, nor moreover is it an asymptote of the curve Dδ, but meets it at some place, such as at λ, although λ can stand apart from here with an infinite interval. That is, the straight line μλ will be indeed *infinite* or bigger than any designated straight line, but not unbounded.
[*The diagram has been simplified*]

6.5 From Leibniz' Marginal Notes to Barrow's Edition of Euclid (1676–1677)

[*To Euclid's definition of parallel lines*]
In general, *parallels* are those lines whose distance is everywhere the same.
[*To Euclid's Parallel Postulate (Axiom 13)*]
This Axiom is difficult. In fact, it is not even true unless the straight lines are in the same plane.

6.6 From *Characteristica geometrica* (January 1677)

His Axiom 11 is: if two straight lines CDF, ABG intersect with a third, EDB, and the angle FDB + GBD is less than two right angles, there will be a point of intersection H on the sides F, G. I am astonished that this is taken as an axiom, since anyone sees how little known it is, and how many difficult things it supposes.

6.7 From *Characteristica geometrica, Scheda 1* (1679)

When two concurrent straight lines are produced they cut each other. This cannot be demonstrated unless the consideration of the plane itself is presupposed. It is necessary to assume, in fact, that two straight lines having a common point are in the same plane, and that, when an indefinite straight line exists in a plane, one part of the plane is on one side of it, and the other part on the other. And that a straight line, which approaches another straight line continuously, also approaches the other side of the plane uniformly and indeed necessarily arrives at the line.

6.8 LH XXXV, I, 11, Bl. 37v. (also in Echeverría 1979, vol. 2, pp. 218-19; slightly corrected)

Duae *lineae parallelae* sunt quae sunt aequidistantes seu inter quas minimae aequales. Hoc definitio locum habere potest in rectis, circulis, parabolis, imo aliis lineis omnibus.

Memorabile theorema
Si duarum parallelarum (secundum definitionem praecedentem) una rectae cuidam ad angulos rectos occurrit, tunc altera eidem etiam ad angulos rectos occurret.

Mirabile paradoxum
Linea parabolae parallela vel ellipsi non est parabola vel ellipsis, sed alterius generis.

6.9 LH XXXV, I, 1, Bl. 13.

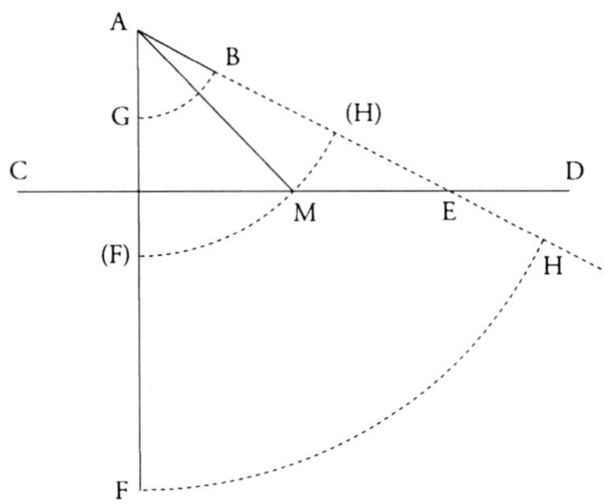

Axioma 13 Euclidis (uti numerat Clavius) demonstratum.

Asserit Euclides rectam AB quae alteri rectae CD propior fit, tandem productam productae occurrere in E.

À puncto A remotiore in recta AB ducatur ad punctum trans rectam interminatam CD situm, F, recta AF, quantum satis est longa quà id punctum F à puncto A quantum satis est remotum esse potest. In recta AF sumatur AF aequalis AG. Centro A radius AGF moveatur donec G incidat in B et F in H, porro autem H esse adhuc trans rectam CD. Quod effici posse sic ostendo. Ponamus H non esse trans rectam CD, sed citra [in (H)], ergo rectae A(F) extremum (F) jam antequam perveniret ad (H) secabit rectam CD in M (nam cum (F) prius esset infra CD, jamque sit supra, necesse est ut alicubi transibit per D, nempe in M). Sed hoc non fecisset, si recta AF sumta fuisset longior quam A(F) vel AM. Cumque idem medium ME <†> possit in quocunque puncto secare, intelligatur antequam G perveniat in B. Sumendo scilicet AF, ab initio majorem, quam est recta ab A ad illud punctum in quo alioqui secasset ducta, hinc patet, effici posse ut non secet AF mota rectam CD, quam postquam G pervenit in B. itaque punctum H est adhuc infra rectam interminatam CD. Ergo recta à puncto supra eam A, ad punctum infra eam H, ducta, AH, necessariò secabit alicubi lineam interminatam seu quantum satis productam CD, in puncto aliquo ut E.

6.8 From *Linea est via puncti* … (1679)

Two *parallel lines* are those which are equidistant, or between which the shortest lines are equal. This definition can apply to straight lines, circles, parabolas, and indeed to all other lines.

Memorable Theorem.
If one of two parallels (following the preceding definition) meets a certain straight line at a right angle then the other also meets the same straight line at a right angle.

Astonishing paradox.
A line parallel to a parabola or to an ellipse is not a parabola or an ellipse, but a curve of another genus.

6.9 *Axioma 13 Euclidis* … (Around 1679)

Euclid's Axiom 13 (as Clavius numbers them) is proved.

Euclid asserts that a straight line AB which becomes closer to another straight line CD, when produced will meet the other, itself produced, at E.

From the remoter point A on the straight line AB, the straight line AF is drawn to the point F, situated beyond the unbounded straight line CD, [*so that AF is*] long enough that point F can be remote enough from point A. In the straight line AF, [AG] is taken to be equal to [AB]. With A as the center, the radius AGF is moved until G meets B and F arrives at H, so that H is by now beyond the straight line CD. That this can be done I show thus: assume that H is not beyond the straight line CD, but above it (at (H)), therefore the extreme (F) of the straight line A(F) will cut the straight line CD at M before it comes through to (H) (for since (F) was earlier below CD, but now is above it, it is necessary that it will have passed through [CD] somewhere, namely in M). But this would not have been done, if the straight line AF had been taken to be longer than A(F) or AM. The same would happen if the middle segment ME is cut, say, before G arrives at B [*? ad sensum*]. Indeed, by taking from the start AF longer than the straight line led from A to the point at which being produced it would cut the other, it is hence obvious, that it can be brought about that AF is moved and does not cut the straight line CD, before G arrives at B. And thus the point H is until that time below the unbounded straight line CD. Therefore the straight line AH is drawn from the point A above CD, to the point H below it, and it will cut with necessity somewhere the unbounded or sufficiently produced CD at some point, namely E.

Sed superest difficultas, nam in demonstratione non explicatur cur requiratur ut AB accedat ad CD, et ideò demonstratio est imperfecta.

Nimirum motus fieri non potest cum recta AF necesse est infinita sit, ad impediendum ne antequam G veniat in D secetur recta CD. Nam ut ex hoc ipso duci potest infinita non movetur, nam tota est <neque> infra nec supra rectam, adeoque eam alicubi secare debet extremo (quo caret) [quanquam haec consequentia ostendenda supersit]. Superest ergo tantum, ut ostendamus hoc contingere unico illo casu, cum recta AB non appropinquat ad CD. Quibus duobus ostensis absoluta erit demonstratio axiomatis.

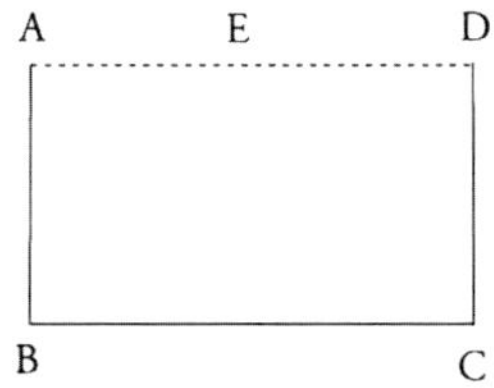

Arabes, Clavius, Guldinus et alij ad hoc axioma Euclidis demonstrandum utuntur hoc principio quod recta quae uno extremo perpendicularis uni rectae est, et super ea ita movetur, altero extremo etiam rectam describit. Hoc illi assumunt velut per se notum, ego sic demonstro: quia unum extremum in una recta movetur alterum in altera et est perpendicularis ab una ad alteram seu minima, erit et vicissim perpendicularis ab ea cui insistit ad eam cui describit. Ergo nulla est ratio earum ad insistere alteri potius describere hanc dicatur. Et vero linea <mota>, ita ut omnia <ejus> puncta rectam describunt.

6.10 Echeverría 1995, pp. 286–94.

Similia voco quae sola comperceptione distingui possunt. Ex quo statim patet, quae similia sunt ea habere homologa (sive respondentia) proportionalia; et contra proportionalia esse homologa sive eodem modo sita. Nam in quibus non sunt partes homologae proportionales, ea distingui possunt per solam memoriam, etsi non ut iuxta se posita spectentur.

…

Recta est cuius pars toti similis est.

…

Parallelae ad quas eandem est perpendicularis.

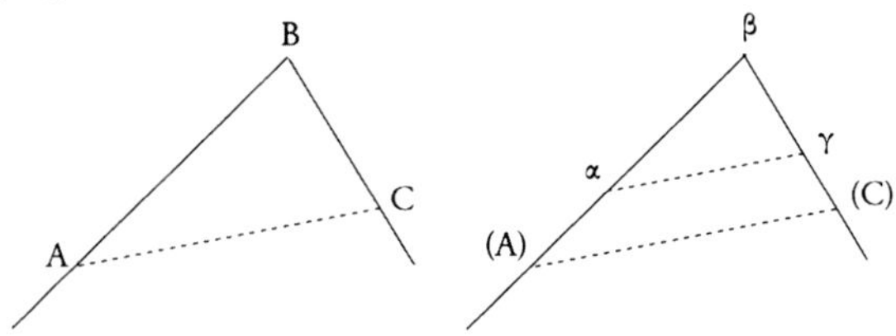

Qui anguli congruunt iidem sunt, demonstrandi ergo in quibus eadem proportio sinuum eos congruere sumatur alius angulus in quo eadem proportio sinuum ajo congruere. Nam si eadem proportio, AB.BC.AC quae αβ.αγ.βγ., ducatur in uno parallela, ut fiant sinus utrobique aequales, (A) (C) parallel. αγ. aequal. AC: quod fieri potest, ergo et (A)β aequal. AB et β(C) aequal. BC ergo congruunt ABC.(A)β(C). Ergo anguli iidem.

But a difficulty remains, namely that in the demonstration it is not explained why it is required that AB approaches CD, and thus the demonstration is imperfect. Indeed, this motion is not possible, since it is necessary that the straight line AF be infinite in order to prevent that the straight line CD is cut before G arrives at B. In fact, precisely from this it can be drawn that an infinite straight line does not move, for (as a whole) it is neither under nor above the straight line [CD], otherwise it should finally cut it somewhere with an extreme, which however it lacks (although this consequence remains to be shown).

Therefore only this remains, that we should show that this [*the meeting of the lines*] happens in the sole case in which the straight line AB does not approach CD. If we show these two things, the demonstration of the Axiom will be absolute.

In order to demonstrate this Axiom of Euclid's, the Arabs, Clavius, Guldin and others used the principle that a straight line which is perpendicular at one extreme to a straight line, and which is moved above it in this way, also describes a straight line with the other extreme. They assumed this to be self evident, but I demonstrate it so: since one extreme is moved in a straight line and the other extreme in the other line, and since it is a perpendicular from one to the other, that is, the shortest line between them, it is also a perpendicular from that on which it stands to that which it describes. Therefore there is no reason for it to be said to stand on the former more than to describe the latter. And indeed the line moves in such a way, that all of its points describe a straight line.

6.10 From *De calculo algebraico et constructiones lineares optime conciliandis* (January 1680)

I call *similar* that which can only be distinguished by coperception. From which it is immediately obvious, that things which are similar have homologous (or corresponding) parts that are proportional; and conversely their proportional parts are homologous or situated in the same way. For those things of which the homologous parts are not proportional can be distinguished through the memory alone; and not only when they are inspected, having been placed near each other.

…

A straight line is that of which a part is similar to the whole.

…

Parallels are those having a common perpendicular.

Congruent angles are identical; it is necessary to demonstrate that angles with proportional sines are congruent. Take another angle whose sines are in the same proportion: I say that it is congruent with the former angle.

For if AB.BC.AC are in the same proportion as $\alpha\beta.\alpha\gamma.\beta\gamma$, and a parallel line is drawn in one of the angles, so that it makes equal sines on both sides, and (A)(C) is parallel to $\alpha\gamma$ and equal to AC (which is possible); then, (A)β is also equal to AB and β(C) equal to BC, and therefore ABC is congruent with (A)β(C). Thus the angles are the same.

Parallelae ad eandem rectam faciunt eundem angulum. Nam quia DC aequ. AE, et eodem jure et AD cum EC. Hinc congruit AEC.CDA. Hinc quia BAC aequ. ACE et ABC aequ. ECF, hinc BAC+ABC aequ. ACF. Ergo trianguli tres anguli 2 rectis aequales, quae demonstratio brevior communis.

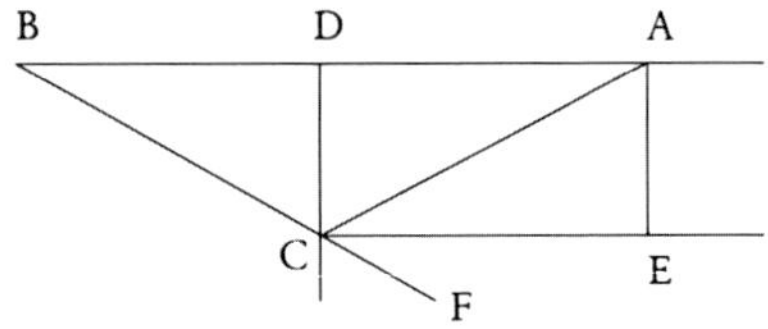

Melius ita omnia demonstramus quam ex Euclidis definitione parallelarum quae imperfecta, quia non ostendit utrum sint possibiles <sive esse> rectae <non> concurrentes. At statim pateri <nostra> esse possibiles, describetur enim una motu puncti ad alteram semper perpendicularis.

6.11 A vi, 4a, n. 124, p. 521.

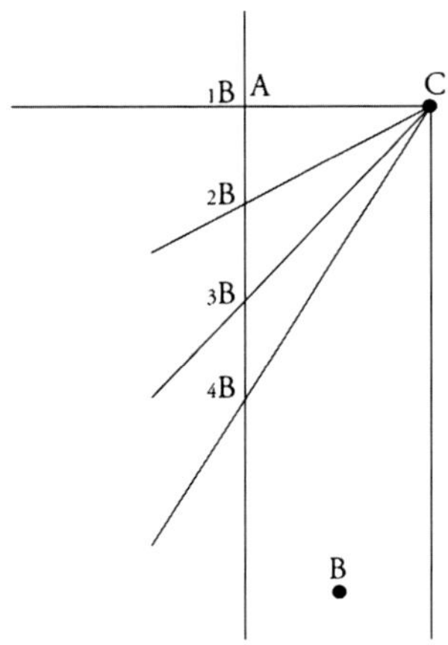

Multum autem interest inter quantitates imaginarias, seu impossibiles per accidens, et impossibiles absolute quae involvunt contradictionem, ut cum invenitur ad problema solvendum opus esse, ut fit 3 *aequ.* 4 quod est absurdum. Imaginariae vero seu per accidens impossibiles, quae scilicet non possunt exhiberi ob defectum sufficientis constitutionis ad intersectionem necessariae, possunt comparari cum Quantitatibus infinitis et infinite parvis, quae eodem modo oriuntur. Nam sit recta indefinita AB transiens per punctum A ad quam sit perpendicularis CA. Et ex puncto C educta secare debet ipsam AB in aliquo puncto B, ut $_1$B, vel $_2$B, vel $_3$B, etc. Patet prout recta CB ad rectam AB angulum facit propiorem recto eo minorem esse AB, adeo ut in casu anguli recti $_1$B incidat in A, seu A$_1$B sit infinite parva sive nulla. Contra quo angulus vel inclinatio rectae CB ad rectam AC magis accedit ad Parallelam eo major erit AB, ita A$_4$B major quam A$_3$B. Et quando recta CB fit omnino parallela ipsi AB, tunc punctum commune B, est imaginarium seu nullum, infinite scilicet hinc distans, et recta AB, est infinita.

(Discrimen tamen inter haec duo imaginaria, quod uno modo problema solubile, altero non; hoc tamen commune, quod saltem per angulum solubile non est, parallelismus enim revera non est angulus.)

Et tamen hujusmodi imaginaria egregium usum habent tum in Conicis, tum alibi passim; ad constructiones universales inveniendas; quod adeo verum est ut saepe calculus ad eas necessario ducat, ubi harum rerum nondum satis periti mire torquentur et in absurdum se incidisse putant. Intelligentes vero sciunt apparentem illam impossibilitatem tantum significare, ut loco rectae angulum quaesitum facientis ducatur parallela; hunc parallelismum esse angulum illum seu quasi angulum quaesitum.

Parallels make the same angle to the same straight line. For since DC is equal to AE and for the same reason AD to EC, AEC is also congruent with CDA. Thus, since BAC is equal to ACE and ABC is equal to ECF, hence BAC+ABC is equal to ACF. Therefore the three angles of the triangle equal two right angles, and this demonstration is shorter than the common one.

We thus demonstrate everything in a better way than from Euclid's definition of parallels, which is imperfect, since it doesn't show whether non-concurrent straight lines are possible (i. e. exist). But it is immediately obvious that our lines are possible, since a parallel is described through the motion of a point always perpendicular to another straight line.

6.11 From *Elementa nova matheseos universalis* (Summer 1683)

There is however a great difference between imaginary quantities, or things impossible by accident, and things impossible absolutely which involve a contradiction, like those that it is necessary to find to solve a problem [*i. e. by a reductio argument*], as when one proves that 3 would be equal to 4 which is absurd.

Indeed imaginary things, or things impossible by accident, which indeed cannot be exhibited because of a lack of the sufficient constitution that is necessary for an intersection, can be compared with infinite and infinitesimal quantities, which arise in the same way. For let AB be an indefinite straight line passing through point A, to which CA is perpendicular. A straight line led out from point C, must cut AB at some point B, such as $_1$B or $_2$B or $_3$B etc. It is obvious that just as straight line CB makes an angle closer to a right angle with straight line AB, so the straight line AB will be smaller, indeed so that in the case in which the angle is right $_1$B meets A, and A$_1$B is infinitely small or nothing. Conversely AB will be bigger the more the angle or the inclination of straight line CB to straight line AB approaches a parallel, so that A$_4$B will be bigger than A$_3$B. And when the straight line CB is entirely parallel to AB, then the common point B is imaginary or nothing, namely standing infinitely apart from here, and the straight line AB is infinite.

(The distinction between these two imaginaries, is that in one of them a problem is solvable, in the other one it is not; but this however they have in common: that the problem is not at any rate solvable through the angle, since strictly speaking parallelism is not an angle.)

However imaginaries of this sort also have an excellent use in conic sections, and in some cases also elsewhere, in finding universal constructions. It is indeed true that calculus often leads to them with necessity, although those who are not skilled enough find them incredible and judge them as leading to absurdity. But those who do understand these things know well that the seeming impossibility only means that a parallel is drawn making a sought angle to a straight line; and thus parallelism is that sought angle, or better a quasi-angle.

6.12 LH XXXV, I, 5, Bl. 18–21. Partially published in Echeverría 1992.

Caetera definitiones satis clarae sunt. Tantum noto ad *Definitionem 29.* dubitari posse an detur quadrilaterum aequilaterum quod simul sit rectangulum, sed si cogitetur rectam rectae aequali ad rectos angulos insistentem in plano moveri quod utique possibile est, manifestum est tale quoque quadratum esse possibile. Idem vitium est definitionis parallelarum, sed si cogitetur finitam rectam super recta indefinita eodem servato angulo moveri, et altero extremo describi lineam manifestum est, eam rectam esse et priori aequidistantem. Rectas autem ambas esse, ex hoc patet quod ad describentem eodem modo se habent.

...

Puto omnia axiomata posse demonstrari, idque majoris esse ad perfectionem Scientiae, quam quis credat.

...

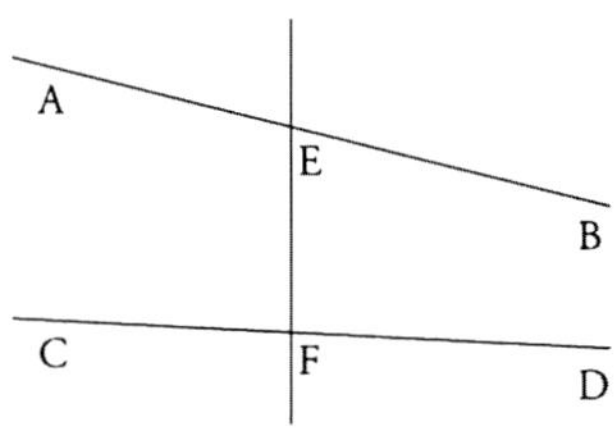

XI. Axioma Euclidis quo Clavio est XIII. demonstrare conatus est Proclus, sed Clavio judice imperfectè, et mihi vicissim Clavij demonstratio non satisfacit. Ita autem habet: si in duas rectas lineas AB, CD altera recta EF incidens internos, ad easdemque partes angulos BEF, DFE duobus rectis minores faciat, duae illae rectae in infinitum productae sibi mutuo incident ad eas partes, ubi sunt anguli duobus rectis minores. Hoc pronuntiato utitur Euclides ad demonstrandum quod si recta incidat in duas parallelas angulos ad eas facit aequales. Nempe angulum AGH angulo CHF. Orta ista perplexitas ex definitione parallelarum, nam si Parallelas definisset Euclides, quae sunt aequidistantes seu minimas distantias ubique habent aequales (cuius corollarium est non posse concurrere, nam in concursu nulla plane distantia est) cessabat difficultas. Ex hac ergo definitione rem ostendam.

...

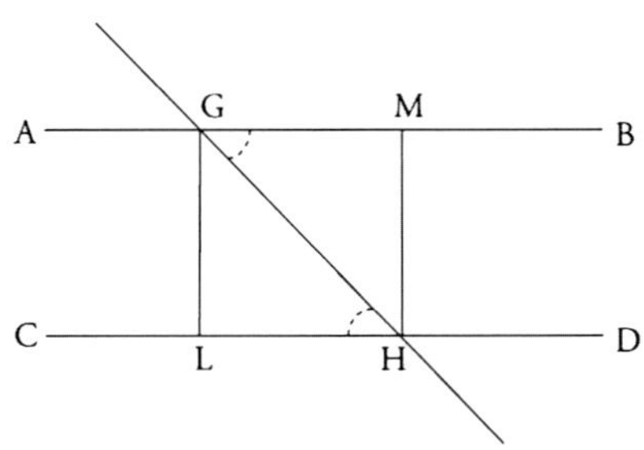

Cum GL et MH sit aequalis moveatur recta LH ubique perpendicularis seu minima, manente angulo recto, ita ut L incedat per LG et H per HM. Utique eodem tempore percurrerentur LG et HM ergo LH et GM sunt aequales. Cum ergo triangula GMH et HLG sint aequalium laterum, erunt ergo et aequalium angulorum, seu angulus GHL aequalis angulo HGM.

6.12 From *Definitiones* (1685)

The other definitions are sufficiently clear. I only note to *Definition* 29 that it could be doubted whether an equilateral quadrilateral could be given which is at the same time a rectangle; but if a straight line is thought to be moved always at right angles to an equal straight line in the plane, which is certainly possible, then it is clear that such a square is also possible. The same fault belongs to the definition of parallels, but if a finite straight line is thought to be moved on an indefinite straight line, while preserving the angle, and a line to be described with the other extreme, then it is obvious, that the latter line is straight and also equidistant to the earlier. It is clear from the preceding that both are straight lines, because they hold themselves in the same way to the describing line.

...

I suppose that every axiom can be demonstrated, and that this is more important for the perfection of science, than it is thought to be.

...

Proclus attempted to demonstrate Axiom 11 of Euclid, which according to Clavius is the 13th, but imperfectly, by Clavius' judgement, and in turn the demonstration of Clavius does not satisfy me.

The axiom is the following: if a straight line EF, making internal angles BEF, DFE with two straight lines AB, CD, on the same sides less than two right angles, when the two straight lines are produced infinitely they intersect with each other on those sides, where the angles are less than two right angles. Euclid used this assertion to demonstrate that if a straight line intersects with two parallels it makes equal angles to them. Namely angle AGH to angle CHF [*reference to a deleted diagram*]. But this difficulty arises from the definition of parallels, for if Euclid had defined Parallels as those which are equidistant or which have equal shortest distances everywhere (it is a corollary of this that they cannot come together, for in an intersection there is clearly no distance at all) the difficulty would stop. Therefore I shall show the matter from this definition.

...

[*deleted passage*]

Let GL and MH be equal, and let the straight line LH, which is perpendicular to them everywhere, or the shortest distance among them, be moved keeping a right angle, so that L proceeds through LG and H through HM. Certainly LG and HM will be completely passed through in the same time, therefore LH and GM are equal. Since therefore the triangles GMH and HLG are equilateral, they will there also be equiangular, or the angle GHL will be equal to the angle HGM.

6.13 LH XXXV, I, 14, Bl. 35.

Si linea recta moveatur vestigiis suis parallela puncta eius quaelibet describunt lineas inter se congruentes.

Si planum ita moveatur ut duae eius rectae (quae productae non coincidant in unam) maneant suis vestigiis parallelae omnes eius rectae manebunt suis vestigiis parallelae, omniaque adeò puncta eius describent lineas parallelas.

Si solidum ita moveatur, ut tres eius rectae maneant suis vestigiis parallelae (vel duo plana) omnes eius rectae (omniaque plana) suis vestigiis parallelae movebuntur, omniaque puncta movebuntur congruè.

6.14 LH XXXV, I, 18, Bl. 14.

Dubitavi diu an duae curvae inaequales (praeter circulos) exacte similes esse possint, ex. g. duae Ellipses, duae parabolae, etc. Notabam enim circulum circulo ubique aequidistantem esse posse, curvam verò aequidistantem Ellipsi vel alteri curvae non esse Ellipsis aut eiusdem cum sua curva denominationis. Et videbatur mihi duas curvas similes et similiter positas necessariò debere esse parallelas, jam verò deprehendi id necesse non esse (si parallelismus pro aequidistantia sumatur esse tunc si parallelismus sumatur ab angulis iisdem quas rectae parallelae ex punctis similiter positis eductae ad ambas curvas faciunt vel eadem recta educta ex centro communi). Et vidi cuilibet curvae aliam similem et similiter positam posse exhiberi an generatione eius ratione simili seu singulatim indiscernibili procedatur.

Ex. gr. si centro C diametro AD describatur semicirculus ABD et quamlibet eius ordinatam ut EB in data ratione producendo usque ad B, sint semielliptica AFD. Et similiter centro eodem C diametro GH describatur alius semicirculus GLH, cuius ordinata quaevis ML in eadem quam dixi ratione producatur in N, fiet semielliptica GNH priori similis et similiter posita, omnia enim singulatim utrobique indiscernibilia sunt.

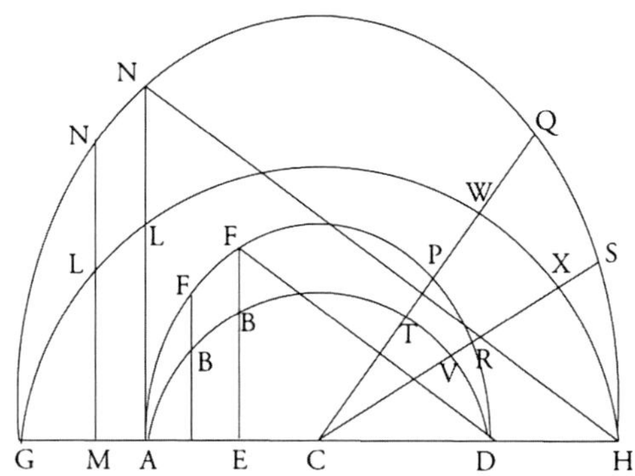

Hinc duae circumferentiae Ellipsive similes erunt inter se ut axis, areae verò ut axium quadrata.

Duae rectae ex centro eductae CPQ, et CRS, duabus ellipsibus FF et NN occurrentes, illi in punctis P et R, huic in punctis Q et S, abscindent sectores et areas similes et similiter positos, eritque sector PCRP ad sectorem QCSQ ut area ellipsium seu ut quadratum AD ad quadratum GH; seu ut sector TCVT respondens circuli generantis minoris ad sectorem respondentem WCXW circuli generantis majoris. Et arcus Ellipticus PR erit ad arcum ellipticum QS, ut arcus circularis PR ad arcum circularem WX, seu ut AD ad GH.

6.13 From *Si linea recta moveatur...* (1685)

If a straight line is moved parallel to its trace then whichever of its points you like describe lines congruent to one another.

If a plane is so moved so that two of its straight lines (which being produced do not coincide in one) remain parallel to their traces, then all of its straight lines remain parallel to their traces and indeed all of its points describe parallel lines.

If a solid is so moved, so that three of its straight lines (or two planes) remain parallel with their traces, all of their straight lines (and all planes) are moved parallel to their traces, and all points which are moved will be congruent.

6.14 From *De curvis similibus et similiter positis et parallelis* (1685)

I doubted just now whether two unequal curves (excluding circles) could be exactly similar, e.g two ellipses, two parabolas, etc. Since I noted that a circle can be equidistant everywhere to a circle, but that a curve equidistant to an ellipse or to another curve is not an ellipse, or a curve of the same denomination as the original curve. And it seemed to me that two curves, both similar and posited similarly, had to be with necessity parallel, but just now I notice indeed that it is not necessary (if parallelism is taken to be through equidistance, or if parallelism is assumed to be from equal angles which two curves make with parallel straight lines led out from similarly posited points or to the same straight line led out from a common center). And I saw that, to an arbitrary curve, another one similar and similarly posited can be exhibited if its generation has a similar, or indiscernible, principle.

E.g. if semicircle ABD is described with center C and diameter AD and an arbitrary ordinate EB is produced with a given ratio to [F], AFD is a semi-ellipse. And similarly another semicircle GLH is described with the same center C and with diameter GH and an arbitrary ordinate ML is produced with the same ratio as I had said before to N, then the semi-ellipse GNH is similar and similarly posited to the prior, since everything is indiscernible considering the figures one by one.

Hence the two similar circumferences or ellipses are among themselves as the axes, and indeed their areas as the square of the axes.

When two straight lines CPQ and CRS are drawn from the center, meeting the two ellipses FF and NN, the first at the points P and R, the second at the points Q and S, they cut sectors and areas that are similar and similarly posited, the sector PCRP will be to the sector QCSQ just as the area of the ellipses or as the square of AD is to the square of GH; or just as the sector TCVT of the corresponding smaller generating circle is to the sector WCXW of the corresponding greater generating circle. And the elliptic arc PR will be to the elliptic arc QS as the circular arc [TV] is to the circular arc WX, or as AD to GH.

Hoc amplius recta quaevis ex centro educta utrique Ellipsi ad eosdem angulos occurrit, seu Ellipses duae in punctis P et Q sunt sibi parallelae seu eandem habent inclinationem, hoc enim sensu hic parallelismum intelligo.

Atque haec quidem si rectae ex puncto quod utrobique eandem functionem facit educantur nempe ex centro communi. Sed si duo sumatur puncta diversa in diversis Ellipsibus eamdemque functionem facientia, ut vertices D et H (ut nunc de focis nil dicam) et ex iis educantur rectae parallelae, ut DF, HN idem ab his rectis parallelis probabitur quod in centro ab eadem ita segmenta DFPD, HNQH erunt etiam inter se ut quadrata axium et arcus eorum ut axes; et ad eosdem angulos curvae occurrent nam et rectae parallelae sunt inter se ut curva ibi parallela curvae nimirum in N eadem inclinatio quae in F, nam et recta CFN ex centro occurret utrique. Idem erit si non ut hactenus rectae, sed alia curva similes et similiter positae ducantur, ea etiam partes circumferentiarum et arearum proportionales abscindunt et ad eosdem angulos occurrent utrobique.

...

Caeterum ita solutus est mihi nodus, cognita vera similitudinis et parallelismi natura, qui me diu vexabat; cum à priori satis intelligerem unamquamque figuram posse contrahi vel expandi in aliam minorem vel majorem per omnia similem et similiter positam, et tamen lineam lineae parallelam ducendo eo sensu ut recta ad rectam perpendicularis etiam esse ad alteram perpendicularis (quo casu aequidistantes sunt) obtinerem curvas planè dissimiles.

6.15 A VI, 4B, nn. 241–242, pp. 1273, 1274, 1281, 1288, 1304.

Parallelam esse lineam lineae resolvitur in *aequalitatem distantiae*, *distantia* in *lineam brevissimam*, seu in *lineam rectam perpendicularem utrique earum quae dicuntur*.

...

1. *rectae inconcurrentes sunt quae continuatae infinite nunquam concurrunt seu se intersecant* 2. *quae eidem rectae sunt parallelae* 3. *quae vel* PARALLELAE *inter se vel* INTER SE DIRECTAE *sunt* (non potuit bona dari definitio parallelarum, quia non consideratum est qualesnam essent inter se directae) 4. *quae in infinitum continuatae vel nihil commune habent, vel prorsus coincidunt*.

...

Rectae
- inconcurrentes
  - parallelae inter se
  - inter se directae
- concurrentes seu inter se vergentes

...

More generally, when a straight line is drawn out from the center and meets both ellipses with the same angles, the two ellipses are parallel to each other at points P and Q, or they have the same inclination, since I understand parallelism here in this sense.

And this indeed holds if straight lines are led out from a point which makes the same function to both, namely from a common center. But if two diverse points are taken making the same function with diverse ellipses, such as vertices D and H (as now I will say nothing about the foci), and from these points parallel straight lines are led out, namely DF, HN, the same will be proved from these parallel straight lines, that was in the case of those from the same center, so that segments DFPD, HNQH will also be among themselves as the square of their axes, and their arcs as their axes; and they meet the curves at the same angles, and the parallel straight lines are among themselves as the curve parallel to the other curve, and without doubt they have the same inclination towards N as towards F, for the straight line CFN also meets both from the center. It will be the same even if not (as until here) straight lines, but other curves both similar and similarly posited are drawn, and they will also cut proportional parts of the circumferences and areas, and meet both lines in the same angles.

…

Moreover in this way I have solved the knot and I have understood the true nature of similarity and parallelism, which had frustrated me for a long time; since I understand well enough, from the preceding reasoning, that any figure can be contracted or expanded into another bigger or smaller figure which is similar in all respects and similarly posited, and nonetheless when I draw a line parallel to a given line in this sense, namely that a straight line perpendicular to one line is perpendicular to the other line (in which case they are equidistant), I obtain curves clearly dissimilar.

6.15 From *Logica de notionibus. Annotata circa schedas Jungianas* (1685)

The notion of a *parallel line* is resolved into the notion of equality of distance to a line, distance is resolved into the shortest line, or the straight line which is perpendicular to both the mentioned lines.

…

1. *Inconcurrent straight lines* are those which infinitely extended never come together or intersect with each other; 2. those which are parallel to the same straight line; 3. those which are either PARALLEL among themselves or DIRECTED toward each other (he cannot give a good definition of parallels, since he did not explain which lines are directed toward each other); 4. those which, extended to infinity, either have nothing in common, or totally coincide.

…

Straight lines
- inconcurrent
  - parallel among themselves
  - directed toward each other
- concurrent, or converging toward each other

…

Rectae largiter *parallelae* dicuntur, *quae in eodem plano sitae nunquam se intersecant, etiamsi in infinitum protendantur*. Eius usus est in theoremate Geometrico de lineamentis paralleliter sitis.

...

Parallelae lineae *werden im Felde gezogen* per angulos oppositos aequales ope octantis vel alterius argumenti dioptrici.

6.16 Robinet 1991, p. 458.

Velim tamen considerari esse aliquid certum etiam in his quae imaginatione non attinguntur; quin et ipsis Geometricis talia immisceri. Talis est notio similitudinis. Et si enim scilicet figuras per angulorum aequalitates definiant Geometrae, ut scilicet rem subito ad imaginem transferant. Facilè tamen intelligitur proprietatem potius, quam naturam figurarum similium sic explicari. Ecce enim et recta rectae similis est, et circulo circulus. Quin et tempori (per se spectato) tempus. Nec lineae inter se similes sunt statim, quae sunt parallelae sive aequidistantes, aut quibus eaedem rectae ad angulos rectos occurrunt. Nam (quod paradoxum videri possit), linea parallela parabolae aut ellipsi, parabola non est, adeoque nec parabolae similis.

6.17 LH XXXV, I, 3, Bl. 1–5.

Demonstrationes Euclideas, ut à Clavio exhibentur revocabimus quoad opus et ratio est, ad calculum situs, quo meliùs talis calculi Elementa constituamus.

...

Axioma XIII. lib. 1 Euclidis numeratum ex ordinatione Clavij suo loco demonstrabimus, quod et Clavius aliò differt.

6.18 LH XXXV, XII, 1, Bl. 201

Parallelae.

Duae lineae parallelae sunt quarum minima intervalla ubique aequalia sunt. Haec definitio etiam lineis curvis applicari potest. Nam accidit, ut recta eadem ubique perpendiculari exsistente; etiam intervallum in perpendiculari illa sumtum, sit semper idem. Quod apparet ex communi generatione duarum parallelarum per evolutionem.

Parallel straight lines are generally said to be those, situated in the same plane, which never intersect, even if they are extended to infinity. This has a use in the geometrical theorem regarding lines situated in a parallel manner.

...

Parallel lines will be *drawn in a field* through equal opposite angles with the help of an octant or another dioptric instrument.

6.16 From *Phoranomus* (1689)

I would like, however, for it to be considered that there is also something certain in those things which cannot be attained by the imagination; indeed such things are mixed in with Geometry itself. Such is the notion of similarity. For instance, if geometers define figures through the equality of angles, they do so in order to immediately transform their subject into an image. Yet, in this way it is easier to understand the property, than to explain the nature of similar figures. And in fact also a straight line is similar to a straight line and a circle to a circle. Indeed even time (considered in itself) is similar to time. Nor are lines always similar among themselves, which are parallel or equidistant, or those which are met at right angles by the same straight lines. For (what can seem to be a paradox), the line parallel to a parabola (or an ellipse) is not a parabola, indeed nor is it similar to a parabola.

6.17 From *Demonstrationes Euclideas* (Around 1690)

We will reduce the Euclidean demonstrations, as are presented by Clavius, to the calculus of situations, insofar as it is useful and necessary; so that we will better constitute the Elements of such a calculus.

...

We will demonstrate in the appropriate place Axiom 13 of Book 1 of Euclid, numbered after Clavius, which Clavius also deferred to another place.

6.18 *Parallelae* (1690)

Parallels.

Two lines are parallel whose minimal intervals are everywhere equal. This definition can also be applied to curved lines. It happens, in fact, that if there is everywhere a straight line which is perpendicular to both curves, the interval of this perpendicular will be always the same. This appears from the common generation of two parallels through evolution.

6.19 GM V, p. 280; GM VII, p. 332.

Et cum constet aliam atque aliam lineam evolutione describi, prout filum producitur longius, animadverteram olim (ut hoc obiter dicam) eas quas Dn. Bernoullius nuper vocavit condescriptas esse *parallelas* inter se, ita ut una sit ab alia ubique aequidistans (seu aequalis ubique minimi intervalli, quod est recta minima ab una ad aliam ducenda) vel ut recta perpendicularis ad unam sit alteri quoque perpendicularis, quae dudum mihi fuit definitio parallelismi in genere sumti.

6.20 LH XXXV, I, 14, Bl. 66.

Omnia sequentia intelligantur esse in eodem plano, etsi non admonetur.

Duae rectae parallelae sunt, quae si termini earum non spectantur, ubique se habent inter se eodem modo. Hoc est ut quemadmodum se res habet si ex unius puncto spectetur altera, eodem modo se res habeat si ex alio spectetur.

Sit una $\bar{P}$ altera $\bar{Q}$ jam qu. P sit A

Si jam $A.\bar{Q} \infty P.\bar{Q}$ erunt $\bar{P}$ et $\bar{Q}$ parallelae

[demonstrandum est, quod $\bar{P}$ et $\bar{Q}$ eodem modo se habeant verum jam video non esse opus si nihil sit in determinantibus, quod ad ipsa sit diversum]

[planum per unamquamque dividitur in duas partes, pars inter duas parallelas intercepta est utrique communis et ita ad ambas eodem modo se habet. Ergo et omnia sic constituta. Idem est etsi non essent parallelae]

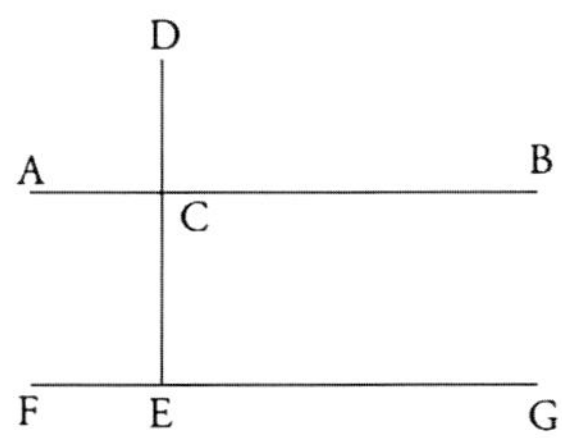

Si sit recta AB, et per ejus punctum C alia recta DE, ubique se habens eodem modo, id est angulo recto, et per hujus punctum E ducatur FEG recta, itidem angulo recto; sed hoc cum fieri posset multis modis, unius est ut simul recta FEG, ubique eodem modo seu congruenter se habeat ad AB; erit utrique parallela. Ex solius rectae DCE motu per rectam AB non describitur parallela, nisi simul intelligamus rectam DE non gyrare nec vacillare inter movendum. Sed hoc non melius <impedietur> nisi simul per 2 parallelas transeat.

6.19 From *Generalia de natura linearum* (*Acta eruditorum*, September 1692)

And since it is established that one line after another are described by evolution, insofar as the string is produced longer, I noticed once (which I mention in passing) that those lines which Mr. Bernoulli recently called "co-described" are *parallel* among themselves, so that one is equidistant everywhere to the other (or the minimal intervals are everywhere equal, which are the shortest straight lines drawn from one to the other), or that a straight line that is perpendicular to one is also perpendicular to the other, which is, since some time, my definition of parallelism in general.

6.20 *Rectae parallelae* (around 1692)

Everything following is understood to be in the same plane, unless cautioned that it is not.

Two parallel straight lines are those which, if their extremes are not inspected, hold themselves everywhere in the same way to one another. This is to say that whatever way the thing holds itself if the one is inspected from a point of the other, the thing holds itself in the same way if seen from the other.

Let one be $\bar{P}$ and the other $\bar{Q}$, and every P be A.

If now $A.\bar{Q} \infty P.\bar{Q}$, then $\bar{P}$ and $\bar{Q}$ will be parallel.

It needs to be demonstrated, that $\bar{P}$ and $\bar{Q}$ indeed now hold themselves in the same way. But I already see that it is not necessary, since there is nothing in the determinations, which would be different among them.

A plane is divided into two parts by each of them, and the part of the plane which is intercepted between the two parallels is common to both and so holds itself in the same way to both. Therefore the same holds also with everything constituted in the same way. It would be the same even if they were not parallel.

If AB is a straight line, and through its point C, DE is led, which holds itself everywhere in the same way, that is at right angles; and through its point E, the straight line FEG is drawn, in the same way at right angles; but this can be done in many ways, one of these being that at the same time the straight line FEG holds itself everywhere in the same way or congruently to AB; it will be everywhere parallel. A parallel is not described from the motion of just the straight line DCE through the straight line AB, unless we understand at the same time that the straight line DE does not rotate or vacillate while it moves. But this is best impeded when it moves through two parallels.

6.21 LH XXXV, I, 12, Bl. 8.

Lineae cuivis in plano positae, ad easdem partes parallelam inscribere.

Hoc fit si circulus in eodem plano super ea intus volvatur, nam centrum eius describet parallelam priori lineam. Circulus tamen non debet justo major, ne centri vestigia se decussent. Potest linea parallela describi per punctum datum si ab eo ducatur minima ad datam, quae erit radius circuli.

Si linea sit in se rediens, videndum an semper linea in ipsa describi possit, similiter in se rediens. Sed hoc adeo fieri non posse, quia evolutione generans limites praescribit; ex qua apparet, quod ex circulo parallelam describente non ita bene intelligitur. Parallelae non sunt similes et similiter positae nisi in figuris regularibus.

Discernuntur res vel per alia (ut duae sphaerae aureae aequales, nempe per situm seu aliorum viciniam), vel per se. Et per se rursus vel per compresentiam (ut duae sphaerae aureae inaequales, si eas simul spectes, vel mensuram ab una ad alteram transferas), vel per se sigillatim, ut sphaeram et ellipsin. Ex hac definitione similium, ut discerni non possint per se sigillatim, sequitur in similibus extensione praeditis esse angulos respondentes aequales, et quantitates respondentes homogeneas esse proportionales, unde sequitur circulos esse ut quadrata, sphaeras ut cubos diametrorum.

6.22 LH XXXV, I, 21, Bl. 5–6. Partially Published in Bopp 1929.

Le livre V des nouveaux Elemens de Geometrie.

Les livres precedens traitoient de la grandeur en generale, ou du nombre indefini. Maintenant commence la geometrie.

Ce qu'on dit de longueur, largeur et profondeur n'explique rien car on n'en a que des idees confuses. De dire que les idees de la droite et de la surface platte sont si simples qu'on ne doit pas leur dessigner une definition n'est pas bien. Car on a de la peine de demonstrer certaines proprietes de la droite et du plan faute de definition.

...

On donne cette definition de la droite qu'elle est la plus courte entre deux points. Mais comment prouver par cette definition des proprietés de la droite par exemple qu'il n'y a qu'une d'un point à u autre au lieu que dans la surface spherique il y a plusieurs minimae d'un point à un autre.

...

6.21 From *Lineam specie datam describere* (Around 1693)

To describe a parallel line on one side of a given arbitrary line in a plane.

This is done if a circle in the same plane is rolled on it, for its center describes a line parallel to the prior. The circle however must not be larger than appropriate, or the trace of the center will cross itself. A parallel line can be described through a given point if the shortest line is drawn from it to the given line, and it will be the radius of the circle.

If a line is returning to itself, it is necessary to see whether a line can always be described in it, similarly returning to itself. But this cannot be done in any case, since the generating line prescribes limits to the evolution; from which it appears, that the description of a parallel with a circle is not well conceived.

Parallels are not similar and similarly posited except in regular figures.

Things can be discerned either through others (as two golden equal spheres, namely through situation or the proximity to others), or through themselves. And again through themselves by compresence (as two unequal golden spheres, if you look at them at the same time, or transfer a measure from one to the other), or through themselves one by one, as a sphere and an ellipse. From this definition of similar things, that cannot be discerned through themselves one by one, it follows that similar extended things have equal corresponding angles, and corresponding homogeneous quantities are proportional; whence it also follows that circles are among them as the squares of the diameters, and spheres are among them as the cubes of the diameters.

6.22 From Leibniz' Notes to Arnauld's *Nouveaux elemens de geometrie* (Around 1693)

The Fifth Book of the New Elements of Geometry.

The preceding books were about magnitude in general, or about indefinite number. Next geometry starts.

That which is said about length, breadth and depth explains nothing, for here we only find confused ideas about them. It is no good to say that the ideas of a straight line and a flat surface are so simple that one must not lay out a definition. For we will have trouble demonstrating certain properties of the straight line and the plane without a definition.

...

The author gives this definition of a straight line: that it is the shortest between two points. But how do you prove, through this definition, the properties of the straight line, for example that there is only one from one point to another whereas on the spherical surface there are several shortest lines from one point to another?

...

Le sixieme axiome que deux droites qui estant prolongées vers un meme coste s'approchent peu à peu se couperont à la fin n'est point vraye si on ne suppose qu'elles soyent dans un meme plan. Et il paroist d'autant plus que cela à besoin de preuve. La raison est encor dans la similitude de la droite car tout ce qui se diminue uniformement, comme icy la distance, evanouit enfin. Hors du plan elle ne se diminue pas uniformement.

...

Sur le livre cinquième des nouveaux Elemens de Geometrie, qui commence à traiter de l'Etendue.

On dit fort bien que l'Estendue est une grandeur continue permanente; mais on se sert incontinent apres des dimensions longeur, largueur et profondeur sans expliquer ce que c'est. J'en ay donné des notions distinctes. L'auetrur ne veut point qu'on definisse le plan et la droite: ce seroit n'en point donner de notion distincte (Et un peu apres il la definit, que c'est la plus courte estendue entre deux points). Il veut aussi qu'on doit commencer par la ligne, mais comme il suppose les lignes dans un même plan, il est manifeste qu'on se sert de la notion du plan ou de la superf.

...

Le second Axiome ou Demande (c'est ainsi que parle l'auteur), ayant deux points donnés on peut mener une ligne droite de l'un à l'autre et l'on n'y en peut mener qu'une; laquelle par consequent est l'unique et naturelle mesure de la distance entre ces deux points. J'y remarque premierement qu'Euclide distingue la Demande de l'Axiome, appellant Demande par rapport aux problemes ce qu'il appelle Axiome par rapport aux Theoremes. Mais icy on joint ensemble l'une et l'autre, car de dire qu'on peut mener une droite d'un point à un autre point c'est une Demande, et de dire qu'on n'en peut mener qu'une c'est un Axiome. Il est vray qu'il seroit à souhaiter que cet Axiome fut demontré et mis en Theoreme. Car dans la surface spherique o peut tirer plusieurs lignes egales entre elles, mais plus courtes de toutes les autres d'un pole à l'autre. Ainsi il n'est pas tousjours necessaire que la plus courte soit unique.

Troisieme Axiome ou Demande. La simplicité de la ligne droite fait qu'en ayant une partie on l'a peut prolonger de part et d'autre jusqu'à l'infini, c'est à dire tant que l'on veut. D'ou il s'ensuite, que deux points estant donnés, toute la ligne droite est donnée. Remarque: la prolongation de la ligne droite à l'infini vient de la similitude de la partie avec le tout. Car puisque sa partie est prolongée dans le tout il faut que ce tout soit prolongé de meme dans un tout plus grand.

Quatrieme Axiome: si une ligne est courbée sur l'autre en une de ses parties elle le sera en toutes pourvu que l'une et l'autre soit prolongée autant qu'il faudra. Rem. C'est que deux droites ne sauroient avoir un segmente commun, comme parle l'Euclide.

The sixth axiom, that two straight lines, which, being extended towards the same side, approach each other little by little, will eventually cut each other is not true at all unless we suppose that they are on the same plane. And it appears all the more that this needs a proof. The reason is again in the similarity of the straight line, for anything that diminishes uniformly, as here distances, eventually vanishes. Outside of the plane the distance doesn't diminish uniformly.

...

On the Fifth Book of the New Elements of Geometry, which begins to deal with Extension.

The author says very well that the Extended is a permanent continuous magnitude; but immediately thereafter he avails himself of the dimensions of length, breadth and depth without explaining what these are. I have given some distinct notions of them. The author really doesn't want us to define the plane and the straight line: this would amount to not giving a distinct conception of them. (And he defines it a little later, as the shortest extension between two points). He also thinks that we must start with the line, but since he supposes that the lines are on the same plane, it is obvious that he helps himself to the notion of a plane or a surface.

...

The Second Axiom or Postulate (that's how the author talks): given two points, one can lead a straight line from the one to the other, and one can lead only one; by consequence this is the unique and natural measure of the distance between the two points. Here, I remark, first, that Euclid distinguished a Postulate from an Axiom, calling a Postulate in relation to problems what he calls Axiom in relation to theorems. But here the author runs the one together with the other, for to say that one can lead a straight line from one point to another point is a Postulate, and to say that one can lead only one is an Axiom. Indeed, it would be desirable that this Axiom be demonstrated and turned into a Theorem. For in the spherical surface one can draw several equal lines, which are also shorter than all the others, from one pole to the other. Hence it is not always necessary that the shortest be unique.

Third Axiom or Postulate. The simplicity of the straight line ensures that having one part of it one can extend it from one side and the other to infinity, that is to say as much as one wants. From which it follows that, with two given points, the whole straight line is given. *Remark*: the extendibility of the straight line to infinity derives from the similarity of the part with the whole. For since its part is extended in the whole it is necessary that this whole be extended in the same way in a bigger whole.

Fourth Axiom: if a line is curved on another in one of its parts it will be so in all its parts provided that the one and the other be extended so long as is necessary. *Remark*: it is because two straight lines couldn't have a common segment, as Euclid says.

Cinquieme Axiome. Deux droites ne se peuvent couper qu'en un point. Remarque: cet axiome est deja compris dans le second.

Sixieme Axiome. Deux droites qui estant prolongées vers un même costé s'approchent peu à peu, se couperont à la fin. Remarque: il faut adjouter, qu'elles sont dans un meme plan. La droite AB coupe le plan indefini א en deux parties congruentes ABC et ABD, et la droite LM en deux autres congruentes LNM et LPM. Si toute la droite (L)(M) tomboit dans une section par AB, comme ABC, il est visible que la section (L)P(M) tomberoit toute dans la section ABC, et donc (L)P(M) seroit congruente avec (L)N(M), c'est à dire avec (L)(M)BA+ADB, donc coupant (L)(M) BA, et l'adjoutant à (L)P(M) il ne peut en provenir ACB, congruant à ADB; c'est pourquoy (L)(M) ne peut tomber dans une section, et par consequent il est commune à deux sections, et ainsi les deux droites se coupent.

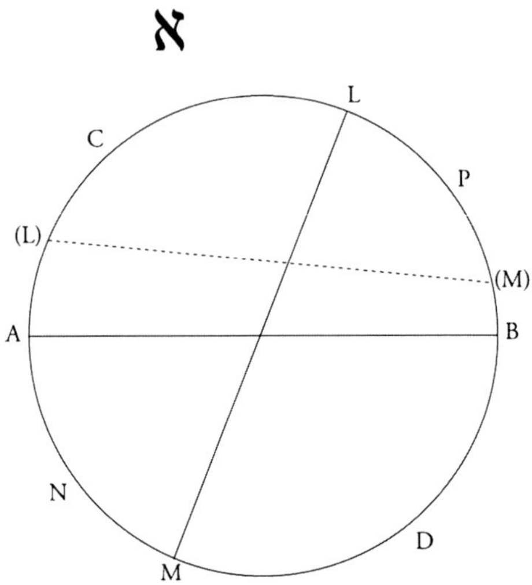

...

Premiere demande sur le cercle, qu'on le puisse decrire ex centro et intervallo datis: pour cela il faut supposer qu'une droite puisse estre mobile dans le plan, une extremité demeurant immobile. *Remarque*. Cecy meme paroist une consequence de l'uniformité del'espace à l'egard de la sphere, et de l'uniformité du plan à l'egard du cercle. Car un point dans un espace ou dans un plan a le meme rapport de tous costés. Mais il merite d'avantage de prouver que tout etendue excepté la droite, et generalement tout corps est mobile, quoyque deux point en luy soyent immobiles, c'est à dire qu'une droite dans l'espace se rapporte egalement de tous costés par rapport à l'espace, lorsque ce qu'on fait se rapporte egalement à elle.

...

Troisieme Axiome. La ligne circulaire est uniforme par tout et par consequent les circonferences d'un intervalle egal sont egales. *Remarque*. Cela suit non pas de l'uniformité du cercle, mais de la nature de la position. Car le centre et le rayon donnés, le cercle est donné. Donc, lors que le points ne different de position il n'y a point de difference entre les cercles, que dans la position.

...

Fifth Axiom. Two straight lines can only cut each at one point. *Remark*: this axiom is already included in the second.

Sixth Axiom. Two straight lines, which, being extended towards the same side, approach each other little by little, will eventually cut each other. *Remark*: it is necessary to add, that they are in the same plane. The straight line AB cuts the indefinite plane ℵ in two congruent parts ABC and ABD, and the straight line LM in two other congruent parts LNM and LPM. If the whole straight line LM fell in one section cut by AB, such as ABC, it can be seen that the section (L)P(M) would fall entirely in the section ABC, and thus (L)P(M) would be congruent with (L)N(M), that is to say with (L)(M)BA+ADB; thus cutting (L)(M)BA, and adding it to (L) P(M), it cannot result that ACB is congruent to ADB. That is why (L)(M) cannot fall in one section, and consequently it is common to two sections, and thus the two straight lines cut each other.

…

First Postulate on the circle, that we can describe it in any given point and with any given radius: to do this, we have to assume that a straight line can be moved in a plane while an extremity of it remains fixed. *Remark.* This seems to me a consequence of the uniformity of space with respect to a sphere, and the uniformity of the plane with respect to a circle. A point in space or plane, in fact, has the same relation with all the sides. But it is worth proving further that each extension except the straight line, and generally each body, is movable even when two of its points are immobile, that's to say that a straight line in space relates equally on all sides with respect to the space, while what one does relates equally to the straight line.

…

Third Axiom. A circular line is uniform throughout and thus the circumferences with the same radius are equal. *Remark.* This does not follow from the uniformity of a circle, but from the nature of position. For with a given center and radius, the circle is given. Thus, since points only differ in position, there is no difference between the circles, other than in position.

6.23 LH XXXV, I, 14, Bl. 108–109.

Duae rectae parallelae sunt, quae ubique eodem modo se invicem habent. Itaque concurrere non possunt, nec sibi reddi viciniores aut remotiores, ergo sunt aequidistantes.

6.24 GM VII, pp. 338–39.

Generaliter autem *parallelas* definio (lineas vel superficies), quarum intervalla (minima scilicet) ubique sunt aequalia. Et cum *rectangulum* (late sumtum) figura sit, quae solos angulos rectos habet, consequens est, figuram hic inter duas parallelas earumque intervalla extrema (minima) contentam ut D(K) merito rectangulum appellari. Ceterum ut hoc quoque adjiciam: *Datae lineae* (curvae vel rectae) *parallelam ducere lineam*, problema est, quod etiam sine evolutione construi potest. Nempe ducatur recta perpendicularis ad lineam datam, et ex puncto ejus aliquo tanquam centro ac pertingente ab eo ad curvam perpendiculari tanquam radio, describatur circulus lineam datam tangens, qui si volvatur super ipsa, describet centro suo lineam datae parallelam. In eo tamen praestat evolutio, quod ejus ope duci potest parallela datae per punctum datum. Tantum enim opus est hoc puncto duci tangentem ad generatricem datae, quae simul et generatrix erit quaesitae, tangente ducta partem fili faciente.

6.25 LH XXXV, I, 1, Bl. 15–19

Expediendi Laboris causa opera pretium erit ut speciosam Situs aggrediamur, supponendo concessa ab aliis, licet demonstratione indigentia. Si rigorose agendum sit quod <tamen> <deinde> suo tempore fiet, cum analysis haec absolvetur. Elementa autem Euclidis à Clavio illustrata hic sequemur speciminis causa.

...

6.23 From *Elementa geometriae generalia* (Around 1694)

Two straight lines are parallel, which hold themselves to each other everywhere in the same way. Thus they cannot come together, nor can they be made closer or remoter, therefore they are equidistant.

6.24 From *De novo usu centri gravitatis* (*Acta eruditorum*, November 1695)

But in general I define *parallels* (lines or surfaces) as those of which the intervals (that is, the shortest lines) are everywhere equal. And since a *rectangle* (in a broad sense) is a figure which has only right angles, the consequence is that this figure is contained between two parallels and their extreme intervals (shortest lines), so that D(K) is deservedly called a rectangle. As another point I will add this as well: the problem *to draw a line parallel to a given line* (curved or straight) can also be constructed without evolution. Namely a perpendicular straight line is led to the given line, and with one of its points as the center and with the part of the perpendicular extending from this point to the curve as the radius, a circle is described, tangent to the given line, which if it is revolved over the given line, describes with its center a line parallel to the given line. The advantage of the evolution, however, is that with its work a parallel to the given line can be drawn through a given point. Since it is only necessary that a tangent be drawn from this point to the curve generating the given curve, which will at the same time also be a generator of the curve being sought, when the tangent takes the role of the string.

6.25 From *Expediendi Laboris causa ...* (Around 1695)

To facilitate our work, it will be useful to prepare a formalism expressing Situation, assuming everything is conceded by others even though it still needs a proof. If it is to be carried on rigorously, this will however come afterwards, in its own time (since this is solved by analysis). To give an essay of this formalism, we will here follow Euclid's *Elements*, illustrated by Clavius.

...

Parallelarum rectarum definitio Euclidea proba est quidem, quod productae sibi nunquam incidant, et tamen in eodem sint plano. Nam extra planum possunt duae rectae dici nec concurrentes nec parallelae. Sed vel hinc intelligi potest ipsam notionem parallelismi aliquod aliud postulare. Nempe parallelarum ubique eandem distantia est, et normalis inter duas parallelas ubique eandem est, quae definitio etiam ad circulos imò et ad alias applicari potest, sed tunc notatu dignum est, si in plano quoque lineam ducas datae curvae parallelam ipsas curvas non fieri ejusdem speciei. Parallelae quoque condefiniri possunt per constructionem curvae quae est locus centrorum osculi. Sed in linea solidi aut lineis duabus non ejusdem plani parallelismus est aliquis non dependens à generatricis osculatoriae descriptione. Caeterum pro rectis parallelis propria dari potest definitio, atque ni fallor optima et facillima, ac minimum supponens; scilicet duae rectae quae ubique se invicem habent eodem modo *parallelae* appellantur. Ex qua definitione non tantum sequitur quae semel à se remotis sumtis semper à se remotas esse, sed etiam caetera quae non sine apparatu Euclidi fere demonstranda.

...

Et revera si quis Euclidea Axiomata neget, ut Franciscum Sanches (autorem libri quod nihil scitur) fecisse accepi, dubitationibus suis ad Clavium missis; fient hypotheses id est hoc saltem habebit Euclides his paucis suppositis concessisque; caetera omnia ab ipso aliisque superstructa fore indubitatae veritates (quodutinam et in philosophia haberemus); sed nunc ipsas illas veritates praeliminares aggrediamur.

...

... nunc ad principia propriè geometrica ad situm pertinentia veniamus.

6.26 XXXV, I, 13, Bl. 15.

Duae rectae *parallelae* sunt, quae productae, si opus, ubique eodem modo se habent.

Tales si sint, ubi sis in illis, ex ipsis discerni nequit.

Sint rectae AE et LR et in illa feratur spectator X, in hac spectator Y, ambo motu aequati uniformi, feranturque ita ut situs ${}_1X_1Y$, et situs tam spectatorum inter se, quam, quoad rectas indefinitas, rectas in quibus ferantur discerni non possit.

Sint adeo omnia non tantum similia semper, sed et aequalia, ut adeò congruentia rectae erunt *parallelae* oportet autem relationem non institui ad aliquod punctum fixum, ut A vel L, alioqui unus utique situs ab alio discerni posset per accessum vel regressum respectu puncti fixi. Itaque si ${}_1X_2X = {}_1Y_2Y$ erit ${}_1X.{}_1Y.\bar{X}.\bar{Y} \simeq {}_2X.{}_2Y.\bar{X}.\bar{Y}$.

...

The Euclidean definition of parallel lines is indeed sound, that being produced they never meet each other, provided that they are in the same plane. For outside the plane two straight lines can be said neither to be concurrent nor parallel. But perhaps from here the notion of parallelism itself can be understood to postulate something else. Namely the distance of parallels is everywhere the same, and the normal between two parallels is everywhere the same, which definition indeed can be applied to circles and other figures; but here it is worth noting, that if in a plane you draw a line parallel to a given curve, these curves will not be of the same species.

Parallels also can be co-defined through the construction of a curve which is the locus of the centers of the osculatory circles. But in a line in space or in two lines not in the same plane parallelism is something not depending on the description of the osculatory generation. On the other hand, a more proper definition can be given for parallel straight lines, and if I am not mistaken, the best and easiest, which presupposes very little; indeed two straight lines which hold themselves everywhere in the same way the one to the other are called *parallel*. From which definition it does not only follow that if they are once assumed to be distant from each other they will always be distant from each other, but indeed other things, which are not to be demonstrated by Euclid without a few complications.

...

And in fact if someone denies the Euclidean Axioms, as I take it Francisco Sánchez (author of the book *Quod nihil scitur*) did, who sent his doubts to Clavius, these should be considered as hypotheses; which means that if only these few suppositions are granted to Euclid, everything else that has been constructed on them by him and others will be indisputable truths (and I would like it if we also had such things in philosophy). But now I will proceed to these preliminary truths.

...

... now we come to the proper geometric principles pertaining to situation.

6.26 From *Duae rectae parallelae* ... (1695)

Two straight lines are *parallel*, which being produced, if necessary, hold themselves in the same way everywhere.

If they are such then, when you are in them, they cannot be discerned from each other.

Let AE and LR be straight lines and an observer X be moved along the first, while an observer Y along the second, both moving with an equal uniform motion, and carried so that the situation of ${}_1X\,{}_1Y$ and both the situations of the observers among themselves, and the straight lines along which they are moved (since they are indefinite straight lines), cannot be discerned.

Indeed if they are not only always similar, but also equal, the congruent straight lines will be *parallel lines*. But it is necessary that this relationship is not referred to some fixed point, such as A or L, otherwise the situation of one can be discerned from the other either through access or regress with respect to the fixed point. Therefore, if ${}_1X{}_2X = {}_1Y{}_2Y$, we have ${}_1X.{}_1Y.\bar{X}.\bar{Y} \simeq {}_2X.{}_2Y.\bar{X}.\bar{Y}$.

...

Axioma: Si similiter determinantia sunt similia, congrua, coincidentia, etiam talia erunt determinata, et vice versa.

Si ab duabus parallelis AB, LM, abscindatur rectae aequales ${}_1X_2X$, ${}_1Y_2Y$, in easdem partes, et duobus rectis ${}_1X_1Y$, ${}_2X_2Y$ jungantur initium initio, finis fini, erunt cunctae rectae inter se aequales.

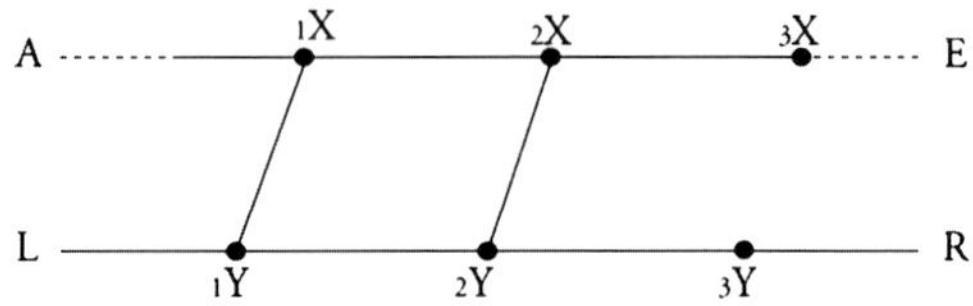

Nam ob ${}_1X.{}_1Y.\bar{X}.\bar{Y} \simeq {}_2X.{}_2Y.\bar{X}.\bar{Y}$ est ${}_1X.{}_1Y \simeq {}_2X.{}_2Y$, adeoque ${}_1X_1Y = {}_2X_2Y$.

Anguli respondentes quas eaedem rectae jungentes faciunt ad parallelas sunt iidem. Nam si ${}_1X_1Y$ alium faceret angulum ad $\bar{X}$, quam ${}_2X_2Y$, etiam (ob axioma) ${}_1X.{}_1Y$ (quae duo puncta determinant rectam ${}_1X_1Y$) aliter se haberent ad $\bar{X}$, quam ${}_2X.{}_2Y$, quod est contra theorema superius, ubi erat ${}_1X.{}_1Y.\bar{X} \simeq {}_2X.{}_2Y.\bar{X}$.

Rectae per ${}_1X$, ${}_1Y$ et per ${}_2X$, ${}_2Y$ sunt parallelae, nam eodem modo se habent inter se ut rectae per ${}_1X$, ${}_2X$, et per ${}_1Y$, ${}_2Y$, seu quadrilaterum habens latera opposita aequalia, habet parallela, vel est parallelogrammum.

Parallelae sunt in eodem plano. Nam iisdem positis, idem est planum ${}_1Y_1X.\bar{X}$ et planum ${}_2Y_2X.\bar{X}$ alioqui situs prior à posteriore distingui posset. Itaque ${}_1Y_1X$ et ${}_2Y_2X$ (parallelae) sunt in eodem plano cum recta $\bar{X}$. Eodem modo ostendetur ${}_1Y_1X$ et ${}_2Y_2X$ esse in eodem plano cum recta $\bar{Y}$, ergo rectae $\bar{X}$ et $\bar{Y}$ sunt in eodem plano.

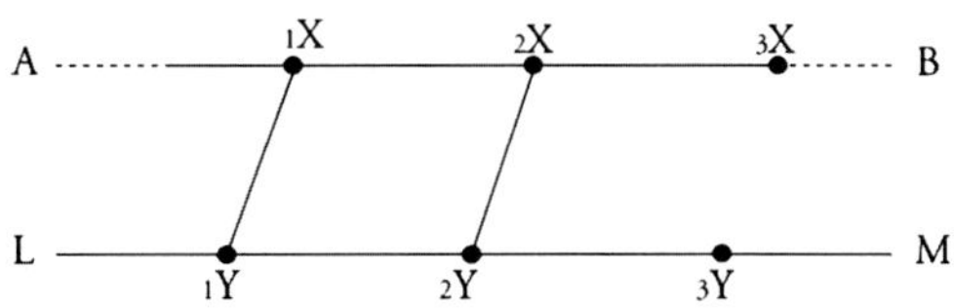

Aliter. Si data sit recta, et extra eam punctum, datum erit planum commune rectae et puncto. Id punctum erit in uno duorum plani segmentorum quae recta facit. Itaque recta parallela per punctum in eodem plano erit, quia quodvis erit punctum ad rectam est, ut unum punctum.

Parallela rectae datae per datum punctum unica est. Quia dato uno puncto ex situ ad quaedam rectae puncta, et segmento plani, datur et aliud ex aliis rectae punctis eodem inter se modo sitis. Datis autem duobus punctis datur recta.

Hoc in calculo exprimetur: sit recta data $\bar{Y}$, punctum datum C, et sit $C.\bar{Y} \simeq X.\bar{Y}$ erit $\bar{X}$ altera recta priori parallela per C.

6.27 A III, 7, n. 72, p. 285.

Et quod duae rectae facientes angulum infinite parvum, habendae sint pro parallelis non quidem absolute admittendum est (cum alioqui etiam rectae quarum concursu determinantur centra circulorum osculantium, non pro concurrentibus sed parallelis haberi deberent).

[*A long section is crossed out*]

Axiom: if similarly determining things are similar, congruent, or coincident; then also their determined things will be such, and vice versa.

If equal straight lines ${}_1X_2X$, ${}_1Y_2Y$ are cut by two parallels, AB, LM on the same sides, and the beginning is joined to the beginning and the end to the end with two straight lines ${}_1X_1Y$, and ${}_2X_2Y$, then all the straight lines will be equal among themselves. For since ${}_1X.{}_1Y.\bar{X}.\bar{Y} \simeq {}_2X.{}_2Y.\bar{X}.\bar{Y}$, then also ${}_1X.{}_1Y \simeq {}_2X.{}_2Y$, and thus ${}_1X_1Y = {}_2X_2Y$.

The corresponding angles, which are made to the parallels when the straight lines are joined together, are equal. For if ${}_1X_1Y$ made a different angle to $\bar{X}$ than ${}_2X_2Y$, then (because of the axiom) ${}_1X.{}_1Y$ (these two points determine the straight line ${}_1X_1Y$) would hold itself to $\bar{X}$ differently from ${}_2X.{}_2Y$, which is against the above theorem, where ${}_1X.{}_1Y.\bar{X} \simeq {}_2X.{}_2Y.\bar{X}$.

The straight lines through ${}_1X$, ${}_1Y$ and through ${}_2X$, ${}_2Y$ are parallel, for they hold themselves in the same way among themselves as the straight lines through ${}_1X$, ${}_2X$, and through ${}_1Y$, ${}_2Y$; or, a quadrilateral having equal opposite sides, has parallel sides, and it is a parallelogram.

Parallels are in the same plane. For, in the previous configuration, the plane ${}_1Y_1X.\bar{X}$ is the same as the plane ${}_2Y_2X.\bar{X}$, otherwise the prior situation could be distinguished from the posterior. Therefore, ${}_1Y_1X$ and ${}_2Y_2X$ (which are parallel) are in the same plane as straight line $\bar{X}$. It is shown in the same way that ${}_1Y_1X$ and ${}_2Y_2X$ are in the same plane as straight line $\bar{Y}$, therefore the straight lines $\bar{X}$ and $\bar{Y}$ are in the same plane.

Another way. If a straight line is given, and a point not on it, a plane common to the straight line and the point will also be given. This point will be in one of two segments of the plane which the straight line makes. Thus, a parallel straight line passing through the point will be in the same plane, since whichever point you like will stand to the straight line, as one point does.

The parallel to a given straight line through a given point is unique. In fact, if one point is given by its situation to the points of a certain straight line, and a segment of a plane is given, then another point is also given among the other points of the straight line situated in the same way among themselves. But, two points being given, a straight line is given.

This is expressed in calculus: let the straight line $\bar{Y}$ be given, and the point C be given, and let $C.\bar{Y} \simeq X.\bar{Y}$, then $\bar{X}$ will be another straight line parallel to the prior through C.

6.27 Leibniz to Johann Bernoulli (January 29th/February 8th, 1697)

And that two straight lines making an infinitely small angle are to be taken as parallels is not indeed absolutely to be admitted – since otherwise also the straight lines whose intersection determines the centers of the osculating circles should be considered not as concurrent lines, but as parallels.

6.28 GM VII, p. 301.

Omnia ergo reduximus ad veritates evidentes, quales sunt: … parallelas ad eandem facere angulos aequales ad easdem partes.

6.29 GM IV, p. 105.

Et cet avantage appliqué encor à la physique et particulierement aux loix du mouvement revient en partie à ce que j'appelle la loy de la Continuité qui me sert depuis longtemps de principe d'invention en physique, et encor d'examen fort commode pour voir si quelques regles qu'on donne vont bien; dont j'avois publié il y a plusieurs années un enchantillon dans les *Nouvelles de la Republique des lettres*, prenant l'egalité pour un cas particulier de l'inegalité et le repos pour un cas particulier du mouvement, et le parallelisme pour un cas de la convergence, etc., supposant non pas que la difference des grandeurs qui deviennent egales est deja rien, mais qu'elle est dans l'acte d'evanouir, et de même du mouvement, qui n'est pas encor rien absolument, mais qu'il est sur le point de l'estre.

6.30 LH XXXV, I, 3, Bl. 13–14.

Attentius examinans quae Proclus et ex eo Clavius attulere ad demonstranda Axiomata ab Euclide Libro Primo sine demonstratione, quae tamen demonstrationem merebantur, assumta; operi paralogismis non carere, quod exponere operae pretium erit, quo magis caveamus in rationibus Geometricis meliusque desiderata aliunde suppleamus; revera enim ad perfectionem scientiae geometricae pertinet haec Axiomata demonstrari, in quo et Apollonium elaborasse ajunt.

…

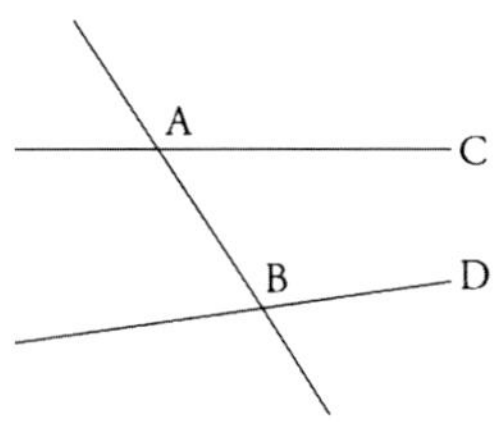

Clavius demonstratum axioma 13 Lib. 1 Euclidis, quod si in duas rectas lineas AC, BD incidens recta AB faciat internos ad easdemque partes angulos BAC, ABD, duobus rectis minores, rectae AC, BD productae ad partes CD, coeant. Assumit tanquam principium per se clarum si linea ubique aequidistet à recta esse rectam, cum et in circulo quod linea ubique aequidistat est circulus. Verum hoc principium in parabola & aliis non procedit, malim ergo hoc principium quod quae aequidistat à recta sit recta, ex eo demonstrari, quod congruat CD linea quae ab extremo F describitur lineae AB quae describitur ab extremitate B.

6.28 From *Specimen analyseos anagogicae* (1698)

Therefore we reduced everything to evident truths, which are such: ... parallels make equal angles on the same sides to the same transversal.

6.29 From *Justification du Calcul des infinitesimales par celuy de l'Algebre ordinaire* (1702)

And this advantage, applied again to physics and particularly to the laws of movement is partially reduced again to what I call the Law of Continuity, which I have used for a long time as a principle of discovery in physics, and moreover as a very convenient principle of examination, to see if some rules which we give work well; I published an essay several years ago about this in the *Nouvelles de la republique des lettres*, taking equality for a particular case of inequality, and rest as a particular case of movement, and parallelism as a particular case of convergence, etc., supposing not that a difference in magnitudes which become equal is already nothing, but rather that it is in the process of vanishing, and similarly for movement, which is not yet absolutely nothing, but which is just about to be.

6.30 From *Attentius examinans ...* (Around 1702)

I am examining the attempts, which Proclus, and after him Clavius, made in order to prove the Axioms that were assumed by Euclid in Book One without demonstration, and that, however, deserved to be demonstrated; they are not free from fallacy, and it will be worth the effort to expose it, since we should take greater care in geometrical reasoning, and we will better supply the desired aims from another source; since to have this Axioms demonstrated indeed pertains to the perfection of the geometrical science, toward which we know that Apollonius also labored.

...

The 13th Axiom of Euclid's Book One was demonstrated by Clavius, that if a straight line AB, intersecting with two straight lines AC, BD, makes internal angles on the same side, BAC, ABD, smaller than two right angles, the straight lines AC, BD, being produced to the sides CD, will come together. However it assumes that the following principle is clear in itself: that if a line is everywhere equidistant to a straight line, then it is itself straight, as happens in the case of a circle, where the line everywhere equidistant is a circle. But this principle does not hold in the case of a parabola or of other lines, and I therefore prefer that this principle, that a line equidistant from a straight line is a straight line, be demonstrated from the fact that the line CD, which is described by the extremity F, is congruent to the line AB, which is described by the extremity [E].

Axioma 13 adhibet Euclides ad demonstrandum prop. 28.1 quod quae anguli ad F et E aequales, quae rectae parallelae, sed idem alias ex eo ostenditur quod ambae utrinque eodem modo EF(F) et FE(E), ergo congruentia. Deinde situs unius alteri congruus, numquam ergo concurrent; rectae sunt ergo parallelae.

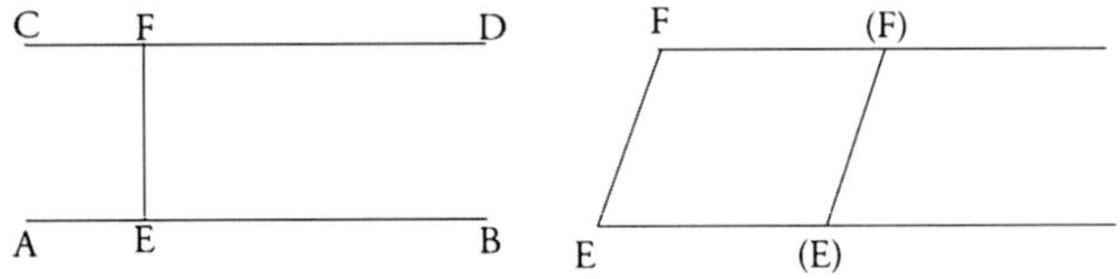

6.31 LH XXXV, I, 3, Bl. 25–32. Partially Published in De Risi 2007.

Calculum Situs haberemus si accommodentur ei Elementa Euclidis.

...

Duas rectas parallelas esse in eodem plano ex hoc colligi potest, parallelae enim sunt rectae quarum concursus infinitè abesse intelligitur. Euclides demonstravit libro XI propos. 2 si duas rectas se secent eas esse in eodem plano. Sed quod rursus sine demonstratione assumsit quod est difficilius, duas parallelas esse in eodem plano: prop. 7 lib. XI. Sed subreptione quadam Euclides hoc definitioni parallelarum inseruit cum parallelas definivit duas rectas <quae> sint in eodem plano, quae productae sibi non occurrant.

Sed si parallelae dicantur rectae aequidistantes, jam ostendendum esset non dari speciem duarum rectarum asymptotarum seu non coincidentium, quae non sint in eodem plano, sunt tamen aequidistantes. Hoc ergo ab Euclide probandum erat. Certe parallelas licet definire nullo ad planum respectu, ut sint aequidistantes. Ostendendum ergo erat ad aequidistantibus rectis convenire definitionem Euclidis, ut sunt in eodem plano. Nam reliqua pars, ut productae utrinque non conveniant, patet ex definitione aequidistantium.

...

Reliquas definitiones libri 1. Euclidis ab 18 inclusivè usque ad finem non est opus ut premam, quia nihil ad calculum situs singulare afferunt, excepta definitione parallelarum; hanc ergo consideram.

Euclid employs Axiom 13 in order to demonstrate *Elements* I, 28, that straight lines which make equal angles to F and E, are parallel straight lines, but the same is proved in another way from the fact that each of the two straight lines EF(F) and FE(E) hold themselves in the same way, and they are therefore congruent. Thus, the situation of one is congruent to the other, and they never come together; therefore the straight lines are parallel.

6.31 From *Calculum situs* (Around 1702)

We will have a Calculus of Situation if the *Elements* of Euclid can be accommodated to it.

...

Two parallel straight lines are in the same plane, since parallel straight lines are understood to be those whose intersection is infinitely distant. Euclid demonstrated in *Elements* XI, 2, that if two straight lines cut each other they are in the same plane. But then he assumed without demonstration the more difficult proposition that two parallels are in the same plane, in *Elements* XI, 7. But with a certain subreption Euclid inserted this into the definition of parallels, since he defined parallels as two straight lines which are in the same plane and being produced do not meet each other. But if parallels are said to be equidistant straight lines, it is now necessary to show that there are no two asymptotic or non-intersecting straight lines, which are not in the same plane, but are still equidistant. Euclid, in fact, had to prove this. Indeed it is admissible to define parallel lines without reference to the plane, as equidistant lines. Therefore it is necessary in order to apply Euclid's definition to equidistant straight lines, to show that they are in the same plane. Because the remaining part, that both being produced they do not come together, is obvious from the definition of equidistance.

...

It is not necessary that I should discuss the remaining definitions of Euclid's Book One, from 18 inclusive up till the end, since they bring nothing remarkable to the calculus of situation, the definition of parallels excepted; therefore I will consider this.

Def. 34, parallelae rectae lineae sunt quae cum in eodem sint plano, et ex utraque parte in infinitum producuntur, in neutrum sibi neutro incidunt. Haec definitio bona est si modo ostendetur coincidere parallelas sic definitas cum iis quae aequidistant. Id manifestum est duas in eodem plano aequidistantes esse parallelas, sed nondum ostensum est duas quascunque rectas aequidistantes esse in eodem plano, sed si coincidentia parallelarum sic definitarum & aequidistantium ostenderetur, confectum esset duas parallelas esse in uno plano. Manifestum est per unum punctum non nisi unam lineam in dato plano duci posse datae rectae in eodem plano aequidistantem, sed quaesitur an absolutè per datum punctum non nisi una recta duci possit datae parallela. Id manifestè constat si adnuisset duas rectas eidem aequidistantes esse aequidistantes inter se, etsi eae tres aequidistantes non sint in eodem plano. Nam sit recta AB, et huic parallela CD per C datum. Dico non posse per C aliam duci ipsi AB parallelam. Nam cum CD et CE sint parallelae eidem AB erunt parallelae inter se, quod est absurdum cum concurrunt in C. Superest ergo ut ostendamus duas AB, EF ad CD aequidistantes, aequidistantes inter se. Hoc Euclides demonstrat 30 primi, sed in plano eodem eas sumens, & per aliam viam.

...

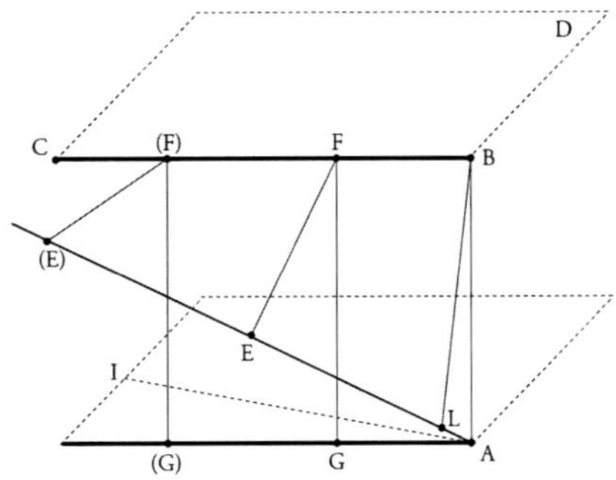

Duas rectas aequidistantes esse in eodem plano tacitè assumsit Euclides, nam etsi parallelas definierit duas rectas in eodem plano non concurrentes, tamen nusquam probavit parallelas quas definivit et aequidistantes rectas (nullo ad idem planum respectu assumtas) easdem esse. Hoc vero cum magni momenti sit, ita suppleo: primum ostendendum per punctum A non nisi unam posse duci rectae datae BC parallelam AG. Ex puncto A in rectam BC agatur normalis AB. Et per ipsam BC agatur planum CBD rectum seu normale ad rectam AB. Et moveatur per BC recta BA cuivis vestigia sint FG, (F)(G) angulo in F semper recto ad planum CBD. Manifestum est rectam AB vel FG semper manere in plano CBA quod planum CBD normaliter secat in recta BC, et eandem rectam FG in eodem plano CBA describere rectam AG. Neque aliam rectam parallelam ipsi BC per A posse a FG describi puncto G dum angulus ipsius FG ad planum CBD manet rectus. Quod si ergo alia recta parallela ipsi BC per A duci potest, oportet ut angulus rectae BA per BC incedentis et puncto B per puncta F(F) transeuntis alicubi ut in F sit obliquus ad planum CBD, adeoque recta quae vestigia repraesentat ipsius BA sit FE, (F)(E). Jam planum BAE secet planum GAI, parallelum ipsi plano CBD, in AI, itaque punctum E cadet intra duo plana parallela. Jungatur AE: cum angulus BAI sit rectus, quia planum GAI est rectum ad BA, et in eodem plano recta AE cadat intra AB et AI, erit angulus BAE acutus. Ex B agatur in AE normalis BL, erit ergo BL minima à B ad AE (quod suppono demonstratum) sed BL minor est quàm BA, ergo BA non est minima inter BC et AE, contra hypothesin. Quia ergo per punctum A non nisi una duci potest aequidistans à BC, quae in eodem sunt plano, ergo ita quascunque aequidistantes esse in eodem plano. In plano autem manifestum aequidistantes nostras et parallelas ab Euclide definitas easdem esse. Ergo absolute sunt easdem.

Def. 34, *parallel lines* are straight lines which, being in the same plane and produced from each side to infinity, meet each other nowhere on either side. This definition is good if there is a way to show that parallels, when so defined, are the same thing as equidistant lines. It is obvious that two straight lines equidistant in the same plane are parallels, but it is not yet shown that any two equidistant straight lines are in the same plane; but if the identity of parallels so defined and equidistant lines is shown, it will come with that that two parallels are in the same plane. It is obvious that through one point only one straight line can be led in a given plane equidistant to a given straight line in the same plane, but it can be asked whether absolutely a unique straight line can be led through this point parallel to a given straight line. This is obviously the case if it is granted that two straight lines equidistant to the same straight line are equidistant to one another, even if these three are not in the same plane. For let AB be a straight line, and CD the parallel to it through the given point C. I say that another straight line cannot be led parallel to AB through C. For since CD and CE are parallel to the same AB they will be parallel among themselves, which is absurd, because they come together in C. Here it remains to be shown that two lines AB, EF, which are equidistant to CD, are equidistant among themselves. Euclid demonstrates this in *Elements* I, 30, but assuming them to be in the same plane, and through another route.

[*A demonstration is crossed out*]

...

Euclid tacitly assumed that two equidistant straight lines are in the same plane, for he defined parallels as two non-concurrent straight lines in the same plane, although he never proved that parallels so defined are the same thing as equidistant straight lines (assuming nothing about the plane). Since this fact is of great moment, I supply it as follows.

First, it is necessary to show that through point A a unique parallel AG to the given straight line BC may be drawn. From point A the normal AB is led to straight line BC. And through this BC the plane CBD is led, at right angles, i. e. normal, to straight line AB. And the straight line BA is moved through BC, whose traces are FG, (F)(G) with the angle in F always right to the plane CBD. It is obvious that the straight line AB or FG will always remain in the plane CBA, that is cut by the plane CBD right angles in the straight line CB; and the same straight line FG, in the same plane CB, describes the straight line AG. Nor can a parallel through A to BC, other than AG, be described from point G, since the angle of FG to the plane CBD remains a right angle. But if another straight line parallel to BC could be led through A, it would be necessary that the angle of the straight line BA proceeding along BC and with point B passing through F(F) somewhere, for example in F, be oblique to the plane CBD, and therefore the straight line which represents the trace of BA is FE, (F)(E). Now the plane BAE cuts plane GAI, parallel to plane CBD, in AI, and thus point E lies between two parallel planes. Let AE be joined. Since BAI is a right angle (for the plane GAI is at right angles with BA), and the straight line AE lies between AB and AI in the same plane, then the angle BAE will be acute. Let the normal BL be drawn from B to AE; this BL will be the shortest line from B to AE (which I suppose demonstrated), but BL is shorter than BA, therefore BA is not the shortest between BC and AE, against the hypothesis. Since therefore a unique line, which is in the same plane, and equidistant to BC, can be led through point A, then it is so that any equidistant lines are in the same plane.

...

Axioma 13 infra suo loco demonstrandum, omnes enim agnoscunt esse theorema.

6.32 C, pp. 439, 477.

Parallela sunt extensa (extrema) aequidistantia, veluti rectae et plana.

...

Parallelum est, quod ubique aeque distat.

6.33 A VI, 6, n. 2, pp. 268, 294–95.

On peut aussi forger des modes impossibles comme lorsqu'on se propose le parallelisme des paraboles en s'imaginant qu'on peut trouver deux paraboles, paralleles l'une à l'autre, comme deux droites, ou deux circles.

...

J'amerois mieux de dire suivant l'usage reçu, que l'Essence de l'or est ce qui le constitue et qui luy donne ces qualités sensibles, qui le font reconnoistre et qui font sa definition *nominale*, au lieu que nous aurions la *definition reelle* et *causale*, si nous pouvions expliquer cette contexture ou constitution interieure. ... Peu de gens ont bien expliqué, en quoy consiste la difference de ces deux definitions, qui doit discerner aussi l'essence et la proprieté. A mon avis cette difference est, que la reelle fait voir la possibilité du defini, et la nominale ne le fait point: la definition de deux *droites paralleles*, qui dit qu'elles sont dans un même plan et ne se rencontrent point quoyqu'on les continue à l'infini, n'est que nominale, car on pourroit douter d'abord si cela est possible. Mais lorsqu'on a compris, qu'on peut mener une droite parallele dans un plan à une droite donnée, pourveu qu'on prenne garde que le pointe du Stile qui decrit la parallele, demeure tousjours egalement distante de la donnée, on voit en même temps que la chose est possible et pourquoy elles ont cette proprieté de ne se rencontrer jamais, qui en fait la definition nominale, mais qui n'est que la marque de la parallelisme, que lorsque les deux lignes sont droites, au lieu que si l'une au moins estoit courbe, elles pourroient estre de nature à ne se pouvoir jamais rencontrer, et cependant elles ne seroient point paralleles pour cela.

But now it is clear that, in a plane, our equidistant straight lines and the parallels defined by Euclid are the same. Therefore they are the same absolutely.

...

Axiom 13 is demonstrated below in its place, since everyone acknowledges that it is a theorem.

6.32 From *Table de définitions* (1702–1704)

Parallels are equidistant extensions (extremes), such as straight lines and planes.

...

Parallel is said of that which stands apart everywhere equally.

6.33 From *Nouveaux Essais sur l'entendement humain*, II, xxxi, § 3; and III, iii, § 18 (1703–1705).

Translated in Remnant&Bennett 1996

But one can also fabricate impossible modes, for example when the parallelism of parabolas is contemplated, through the delusion that two parabolas can be found which are parallel to one another, like two straight lines or two circles.

...

I would sooner say, in keeping with accepted usage, that the essence of gold is what constitutes it and gives it the sensible qualities which let us recognize it and which make its *nominal* definition; whereas if we could explain this structure or inner constitution we would posses the *real*, *causal* definition. ... Not many people have properly explained the difference between these two definitions, a difference which also marks off essence from property. In my opinion, the difference is that the real definition displays the possibility of the *definiendum* and the nominal does not. For instance, the definition of two *parallel straight lines* as "lines in the same plane which do not meet even if extended to infinity" is only nominal, for one could at first question whether that is possible. But once we understand that we can draw a straight line in a plane, parallel to a given straight line, by ensuring that the point of the stylus drawing the parallel line remains at the same distance from the given line, we can see at once that the thing is possible, and why the lines have the property of never meeting, which is their nominal definition (though this is a sign of parallelism only when both lines are straight: if at least one were curved they might be such that they could never meet, without thereby being parallel).

6.34 GM VII, pp. 342–43; GM III, pp. 793–94.

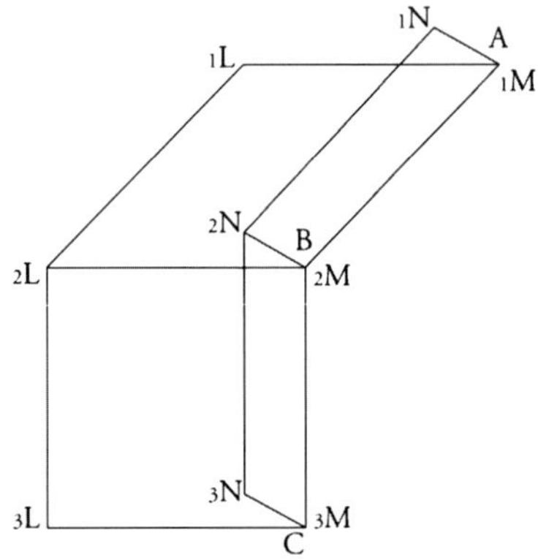

Porro curva, quae basin nonnisi radit adeoque vestigiis suis parallela incedit, singulis punctis plani sui mobilis describet lineas basi similes et aequales seu congruentes. Nam rectae ${}_1N_2N$, ${}_2N_3N$ respectivae parallelae et aequales sunt ipsis ${}_1M_2M$, ${}_2M_3M$, id est rectis AB, BC, idemque est de motu puncti L, quod de motu punctorum M et N. Et si quis agnoscat, quaevis plani mobili puncta ut L et N describere lineas congruas inter se, idem ei dicendum erit de puncto sustentante M, nempe describere lineas prioribus congruas, cum punctum N ipsi inassegnabiliter vicinum assumi possit, adeoque in ipsum tandem incidere intelligatur. Quia autem M describit basin, utique lineae a punctis L et M descriptae basi congruent. ... Si basis esset circulus, figura super basi incedens radendo moveretur eo motu, quem Hobbius olim vocavit circularem simplicem et de eo quo quaedam ingeniose jam tum observavit in *Elementis de Corpore* et in *Problematibus*, et inter alia, quod dum quaelibet recta in plano sic moto designata manet suis vestigiis parallela, necessario durante motu quaevis puncta plani describant lineas inter se aequales et similes. Sed nos addimus, eundem esse hunc motum cum motu curvae aliam uno puncto suo radentis.

...

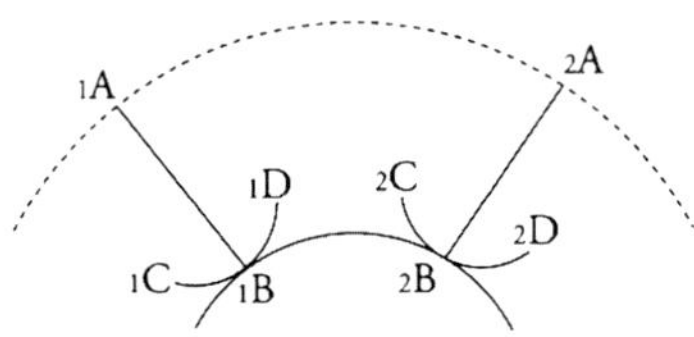

Vereor ne jure quaedam in mea motus radentis explicatione desideres; cum enim nuper tandem nonnihil animadverterem, statim deprehendere mihi videbar, curva aliam curvam mere radente omnino mutari parallelismus rectarum, et nisi fallor, punctum quodvis determinatum rectae AB ad ambas curvas BB et CDB normalis, in mobili curvae CBD plano ductae et cum eo procedentis, describere lineam AA, basi BB equidistantem. ... neque omittam aliquando rem retractare in Actis.

6.35 De Risi 2007, pp. 601–602.

Demonstratio omnimoda quae nullam propositionem, nisi identicam, sine demonstratione assumit; nulloque proinde axiomate aut alia assumtione demonstrabili sed non demonstrata utitur; etiam analysin perficit.

Itaque ut Analysin Situs constituerem, inter alia cogitavi scopum obtineri posse si perficerentur demonstrationes Elementorum Euclidis, in quo jam olim Apollonius, Proclus, nuper Clavius laboraverunt.

...

6.34 From *De lineae super linea incessu* (*Acta eruditorum*, January 1706), and Letter to Johann Bernoulli (July 15th, 1706)

[A] Moreover a curve which touches only the base and then proceeds parallel to its trace describes with the single points of the movable plane lines similar and equal, i. e. congruent, to the base. For the straight lines $_1N_2N$, $_2N_3N$ are respectively parallel and equal to $_1M_2M$, $_2M_3M$, that is to the straight lines AB, BC, and it is the same with the motion of point L, as with the motion of points M and N. And if someone recognizes that, whatever points in a movable plane, such as L and N, describe lines congruent to one another, he must also say the same thing about the base point M, namely that it describes lines congruent to the prior lines, since point N can be assumed to be unassignably near to it, and indeed it can be understood finally to meet it. Since however M describes a base, the lines described by points L and M are also congruent to the base.

...

If the base were a circle, the figure moving along the base in radent motion would have that kind of motion which Hobbes once called *simply circular* and regarding which he made a few ingenious observations in the *De corpore* and in the *Problems*; among other things, he observed that when an arbitrary straight line in a plane, endowed with a such a motion, remains parallel to its trace, it is necessary that in the motion any point on the plane describe lines equal and similar to one another. But we add ourselves, that this motion is the same as the motion of a curve which is radent to another curve with one of its points.

...

[B] I fear that you will now rightly desire my explication of radent motion; since finally just now I considered it a bit, it seems to me that I realized at once, that if a curve only moves with a radent motion along another curve, the parallelism of straight lines is completely changed, and unless I am mistaken, an arbitrary point of the straight line AB, normal to both curves BB and CBD, which is in the moving plane of the curve CBD and proceeding with it, describes a line AA, equidistant to the base BB.

...

... and I won't omit to retract my views, at some time, in the *Acta Eruditorum*.

6.35 From *Demonstratio omnimoda ...* (1712)

A thoroughgoing demonstration is one which assumes no proposition, except identity, without demonstration; and it uses no prior axioms or other undemonstrated but demonstrable assumptions; and it also perfects analysis.

And thus in order to constitute an Analysis of Situation, I thought among other things that the aim could be obtained if the demonstrations of the *Elements* of Euclid were perfected, towards which already, long ago, Apollonius, Proclus and more recently Clavius have worked.

...

Euclides quoque assumit hoc axioma quod apud Clavium est 13: si in duas rectas altera recta incidens internos ad easdemque partes angulos duobus rectis minores faciat, duae illae rectae in infinitum productae sibi mutuo incident ad eas partes ubi sunt anguli duobus rectis minores. Hoc pronuntiatum Geminus, Proclus aliique à numero principiorum rejiciunt. Proclus demonstrare conatus est sed assumit duas rectas ex eodem puncto eductas continuata productione ita à se invicem recedere, ut distantia earum sit major data. Clavius aliter rem demonstrat et satis quidem prolixè assumendo lineam cujus omnia puncta à recta linea, quae in eodem cum ipsa plano existit, aequidistant, rectam esse. Clavius audiit axiomatis Euclidei demonstrationem quandam extare in Euclide quodam arabico, sed negat se ejus communicationem à libri possessore impetrare potuisse.

6.36 GM v, pp. 200–204, 208–211, with some Corrections from the Manuscript in LH XXXV, I, 1, 1–10.

XXXIV. *Parallelae rectae* lineae sunt, quae cum in eodem modo sint plano, ex utraque parte in infinitum productae in neutram sibi mutuo incident.

(1) Haec definitio videtur parallelae magis ex proprietate remotiore, quam natura apertiore describere, et dubitare poterit aliquis an dentur, seu an non omnes rectae in eodem plano tandem conveniant. Demonstratio autem, quae ostendet tales rectas dari posse, naturam earum aperiet ex aliquo priore, de cujus possibilitate dubitari non possit.

(2) Euclidem haec definitio coëgit assumere Axioma, quod in Clavii editione est 13^{mum}, sed quod Veteres jam fassi sunt demonstratione indigere, nec bene locum inter Axiomata tueri.

(3) *Parallelae* possunt definiri rectae, quae se invicem ubique habent eodem modo. Tales notiones possibilitatem suam secum ferunt, nam cum spatium sit ubique uniforme, et recta non minus, manifestum est, punctum, quem situm ad rectam aliquam habet, eundem posse repetere, atque ita in motu eundem posse servare, atque eo vestigium motus seu lineam, quam describat, semper se eodem modo ad eandem rectam habere. Dupliciter autem punctum in motu existens situm ad rectam servare potest, tum eundem servet ad eadem puncta rectae, ut cum punctum circa axem gyratur, tum etiam cum variat situm ad puncta, sed eodem modo se habet ad nova, adeoque situm ad rectam servat, etsi ad puncta non servet.

Euclid also assumed this axiom which according to Clavius is the 13th: if a straight line intersects with two straight lines and makes angles on the same side less than two right angles, the two straight lines being produced to infinity will intersect each other mutually on the side where the angles are less than two right angles.

Geminus, Proclus and others excluded this assertion from the number of principles. Proclus tried to demonstrate it but assumed that if two straight lines are led out from the same point with a continuous production, the distance between them will become greater than any given distance. Clavius demonstrated the matter in another way and with a certain prolixity by assuming that a line whose points were all equidistant from a straight line, which is in the same plane, is a straight line. Clavius heard that a demonstration of Euclid's axiom existed in a certain Arabic Euclid, but denied that he could have obtained an account of it from the possessor of the book.

6.36 From *In Euclidis* πρῶτα (1712)

Definition 34. *Parallels* are straight lines, which, being in the same plane, when produced infinitely in both directions, do not intersect with each other mutually in either direction.

(1) This definition seems more to describe parallels from a remoter property than from a clearer nature; and someone could doubt that parallel lines may be given, or that all straight lines on the same plane don't eventually come together. However, a demonstration that shows that such straight lines can be given, would make their nature clear from something prior, the possibility of which could not be doubted.

(2) This definition drove Euclid to assume an axiom, which is the 13th in Clavius' edition, but which the ancients already granted needed a proof, and that it was no good to defend its place among the axioms.

(3) *Parallels* can be defined as straight lines which hold themselves in the same way to each other everywhere. Such notions bring their possibility with themselves, for, since space is uniform everywhere, and straight lines no less, it is obvious that, a point, which has a situation to some straight line, can repeat this situation, and thus can preserve it in motion, and so the trace of this motion, or the line which the motion describes, will always relate in the same way to the same straight line. There are however two ways in which a point existing in motion can preserve its situation relative to a straight line: in one way, it can preserve the same situation to the same points on the straight line, so that the point rotates around an axis; but in another way, while its situation with respect to the points changes, it nonetheless holds itself in the same way each time to new points, and so completely maintains its situation relative to the straight line, although it does not preserve its situation relative to the points.

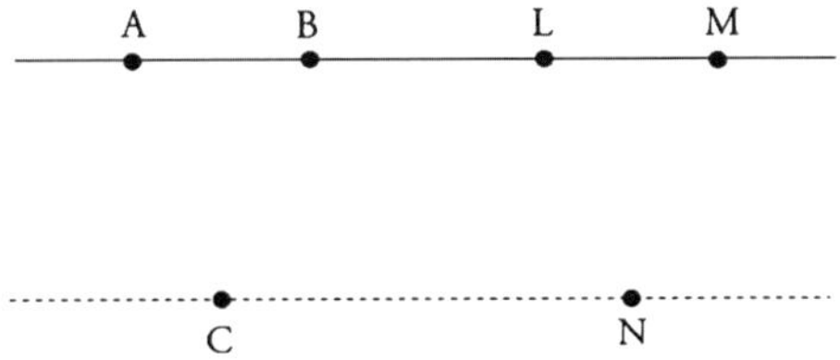

(4) Posset parallela ad datam sic determinari. Ex punctis A et B, in recta sumtis, determinetur punctum C, et ex punctis L et M, eodem modo sitis inter se, (posita AB = LM) determinetur eodem modo punctum N, puncta C et N eodem modo se habebunt ad rectam indefinitam per A, B; L, M. Cum ut A, B, ita se habeant in recta ut L, M itaque C et N sunt in parallela ad rectam, et quidem erunt etiam in recta, quia recta determinata, transiens per C et N, eodem modo etiam se ad rectam ubique habeat. Sed eodem modo determinata quaevis alia puncta, quotcunque aliis ut A, B vel L, M assumtis, dabunt puncta in parallelam cadentia, quae quidem omnia cadere in rectam, etiam ex eo intelligi potest, quia ex C et N recta determinatur.

(5) Modi determinandi varii intelligi possunt. Ex. gr. si plani considerationem seponamus, potest concipi circa puncta A et B describi sphaeras, quarum superficies sese secent. Sectio duarum superficierum sphaericarum erit circulus. Hujus centrum potest esse punctum C. Sed si planum adhibeamus, possumus concipere in plano describi circa A et B circulos, quorum intersectio ab uno latere rectae est in puncto. Eodem modo L et M ab eodem latere dabunt punctum N.

(6) Sed quia elegimus, rectam considerare ut plani sectionem, ponamus in eodem plano moveri rectam mobilem, quae durante motu eundem semper situm servet ad rectam quiescentem, eundem nempe angulum, linea utique, quam describet, etiam ubique se eodem modo ad rectam habebit. Et commodissimum erit assumi angulum rectum.

(7) Sed ostendendum foret, lineam quae describitur etiam esse rectam. Colligemus autem ex eo, quia ad eam, quae ubique uniformis est, se habet ubique eodem modo et ipsam esse ubique uniformem.

(8) Concipi etiam possunt parallelae sine ulla mentione recti, etsi ex eo conceptu sequatur esse rectas. Ponatur duo puncta A et B moveri motu aequiveloce, ita ut vestigia eorum a se invicem discerni non possint, et eandem etiam semper habeant relationem inter se, et has lineas vocari parallelas. Talis definitio non nisi parallelis rectis quadrat. Nam etsi puncta, circulos concentricos describentia eadem celeritate aequali, etiam eundem semper situm servent, una circumferentia ab alia discerni potest.

(9) Possunt etiam parallelae definiri rectae aequidistantes, seu tales ut minima, ab una ad aliam ducta, sit ubique aequalis. Sed de minimis a recta ad rectam nondum actum est.

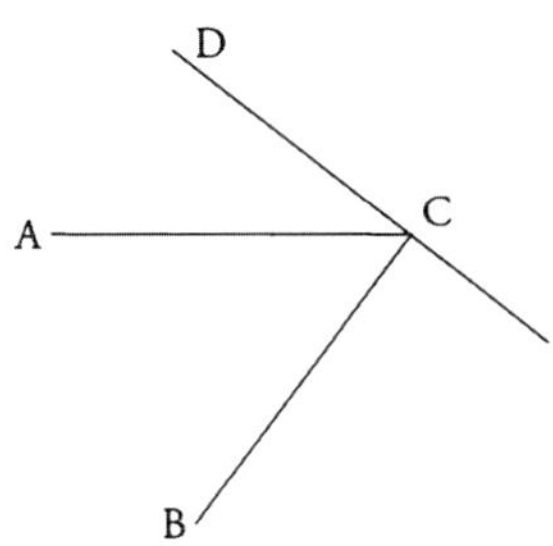

(10) An ergo parallelas definiemus commodius duas rectas ejusdem plani, quae ad eandem rectam faciunt angulum eundem, quod si eas ponamus hoc modo se habere ad quamcunque rectam. Hinc statim sequitur eas non posse concurrere. Concurrant enim AC et BC in puncto C, dico non esse parallelas quia recta DC, quae iis occurrit in C, non potest angulum aequalem facere ad ambas. Nam si aequales essent anguli DCA et DCB, aequaretur pars toti.

(4) A parallel to a given straight line may be thus determined. From points A and B, lying on a straight line, point C is determined, and from points L and M, fixed with the same situation among themselves (posit AB = LM), point N is determined in the same way, so points C and N hold themselves in the same way to the indefinite straight line through AB and LM. Since A, B hold themselves to each other in the same way on the straight line as L and M, thus C and N are in a line parallel to the straight line, and they are indeed in a straight line, since the determined straight line, which passes through C and N, indeed holds itself in the same way everywhere to the straight line. But as many other points as you like are determined in the same way: whenever different points are taken for A, B or L, M, they give points lying in a parallel, which moreover all lie in a straight line. This can be understood from the preceding, since a straight line is determined by C and N.

(5) Various ways of determining can be understood. For example, if we set aside the consideration of the plane, it is possible to conceive spheres described around points A and B, whose surfaces intersect. The section of the surfaces of the two spheres will be a circle. The center of this can be point C. But if we use a plane, we can conceive circles described in a plane around A and B whose intersection on one side of the straight line is in a point. In the same way, L and M give point N on the same side.

(6) But since we choose to consider the straight line as a section of a plane, we posit that a mobile straight line is moved in the same plane, which during motion always preserves the same situation to a resting straight line, namely the same angle, so that the line, which it describes, will always hold itself in the same way to the straight line. And it will be most convenient to assume that this angle is a right angle.

(7) But it will be necessary to show that the line which is described is indeed a straight line. But we gather from the preceding that, since it holds itself in the same way everywhere to it [*the straight line*], which is uniform everywhere, it is itself also uniform everywhere.

(8) Parallel lines can be conceived without any mention of straightness, although from this conception it follows that they are straight lines. It is posited that two points A and B are moved with a motion of equal speed, so that their traces cannot be discerned the one from the other by themselves, and so that they also always have the same relation to one another: let these lines be called parallels. Such a definition only squares with parallel straight lines. For although points describing concentric circles with the same equal speed also always preserve the same situation, one circumference can be discerned from the other.

(9) Parallels can indeed be defined as equidistant straight lines, or as those that are such that the shortest lines drawn from one to the other are everywhere equal. But we have not yet treated the topic of the shortest lines from a straight line to another.

(10) Or therefore we can define parallels more conveniently as two straight lines on the same plane, which make the same angle to the same straight line, but only if we posit that they will hold themselves in this way to any straight line. Hence it follows immediately that they cannot meet. Since if AC and BC meet in point C, I say that they are not parallels since the straight line DC, which meets with them at C, cannot make an equal angle to both. For if the angles DCA and DCB were equal, the part would be equal to the whole.

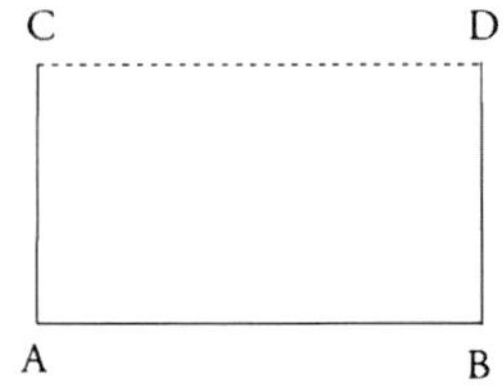

...

(12) Sed ostendere oportet, si una recta ad duas aequales angulos facit, quamvis ad eam angulos facere. Multa tentavi, et video ne hoc quidem facile demonstrari posse, si ex recta AB educatur perpendiculares aequales AC et BD, esse CD rectam aequalem ipsi AB, et angulos ad C et D esse rectos. Quemadmodum ponendo rectam ex C versus D angulo recto educi, non facile demonstrari potest incidere in D et angulo itidem recto. Euclides cum difficultatem in proprietatibus parallelarum demonstrandis reperiret, axioma assumsit 13, quod rectae ex recta eductae, angulis inaequalibus (eo enim res redit) concurrant adeoque non sint parallelae, nam has definivit, quae in plano non concurrunt. Si maluisset definire parallelas, quae aequales angulos faciunt ad rectam, assumendum fuisset ei axioma, quod tales non concurrant. Itaque res eo redit, ut connexio inter haec duo demonstretur, aequalem angulum facere ad rectam, et non concurrere. Demonstravi paulo ante, si ponatur angulum aequalem fieri ad quamcunque rectam. Ergo hoc ostendendum erat.

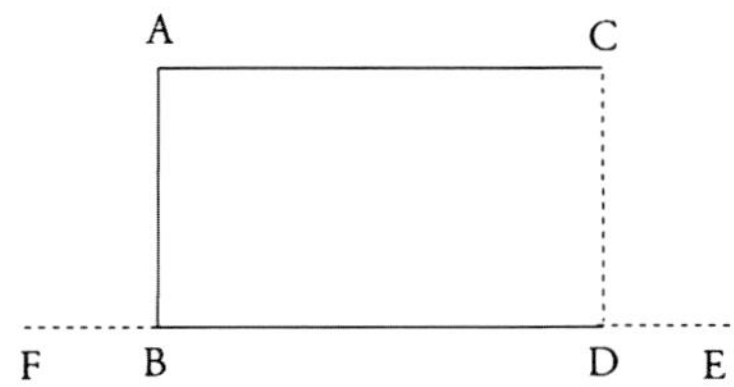

(13) Rem omnem consequi posse mihi videor ex altiore principio, nempe rationis determinantis.

Ex A, B angulo recto eductae AC, BD sunt determinatae. Assumantur AC, BD aequales. Ubique ob determinata puncta C et D, determinata est recta DC. Eritque angulus ad C aequalis angulo ad D, cum eodem modo determinentur. Quin eodem argumento sequitur, angulum ad D vel ad B esse rectum, quia ex determinatione nulla haberi potest ratio, cur CD se aliter habeat ad partes E quam ad partes B, vel cur AB se aliter habeat ad partes D quam ad partes F nec ullum est principium determinandi anguli obliquitatem. Itaque anguli ABD et ABF sunt aequales, atque adeo AB est perpendicularis.

(14) Ex his patet congrua esse LACN, MBDO et NAL, OBM, angulis A, B, C, D rectis, sive inverti posse positio LA=CN, et BM=DO, ergo congruent etiam rectae LM et NO. Sed LM recta ipsis AL, BM est in directum, ergo NO, aequalis ipsi LM, etiam est ipsis CN, DO, seu CNDO est recta.

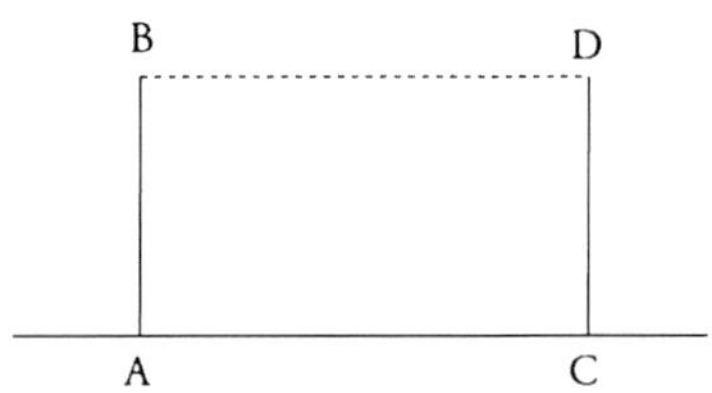

(15) Eodem modo ponendo AB procedere super AC angulo recto servato in CD, nulla ratio esse potest cur angulus DBA et DBC non sint aequales, seu cur plus aut tanto plus inclinetur in alteram partem. Unde patet etiam, si CD=AB et angulus ad A et C rectus, fore et ad B et D rectum. Unde simul probatur, et BD rectam esse, ea enim sola utrinque <se> habet eodem modo.

[*There are two deleted versions of § 11 and four of § 12 here*]

(12) But it is necessary to show, if one straight line makes equal angles to two, then any straight line would make equal angles to them. I have tried many things and I see that this indeed cannot be easily demonstrated, that if from the straight line AB the equal perpendicular lines AC and BD are drawn out, then CD is a straight line equal to AB, and the angles in C and D are right angles. Even though we assume that the straight line from C towards D is drawn out with a right angle, it cannot easily be shown to intersect with D with a right angle there. Since Euclid met a difficulty in demonstrating the properties of parallels, he assumed axiom 13, that when straight lines are drawn out from a straight line, with unequal angles (since the argument reduces to this), they come together and thus are not parallels; for he defined parallels, as those which do not come together in a plane. If he had chosen rather to define parallels, as those which make equal angles to a straight line, then it would have been necessary for him to assume this axiom: that such lines do not come together. And indeed the point is that the connection between these two should be demonstrated: making equal angles to a straight line, and not coming together. I demonstrated this a little before, if we assume that an equal angle is made to any straight line. Therefore it was necessary to show the latter assumption.

(13) It seems to me that the whole thing can follow from a higher principle, namely that of Determinant Reason. The straight lines AC, BD, drawn out with a right angle from A, B, are determined. Assume AC and BD are equal. The straight line DC is determined since the points C and D are determined. And the angle to C will be equal to the angle to D, since they were determined in the same way. Verily it follows with the same argument, that the angle to D or to B is a right angle, since from the previous determination no reason may be found, why CD should hold itself differently to side E than to side B, or why AB should hold itself differently to side D than to side F, and nor is there any principle of determining the inclination of the angle. And so angles ABD and ABF are equal, and thus AB is perpendicular.

(14) It is obvious from these that LACN, MBDO and NAL, OBM, are congruent, with angles A, B, C, D being right angles, and thus the positions may be inverted, so that LA=CN and BM=DO, and therefore also the straight lines LM and NO are congruent. But the straight line LM is on a straight line with AL, BM, and since NO is equal to LM, also DO is on a straight line with CN, that is, CNDO is a straight line.

(15) In the same way, if AB is posited to proceed on AC, preserving a right angle to CD, there can be no reason why angles DBA and DBC are not equal, or why one of them should be inclined, or more inclined than the other, toward one side. Whence it is also obvious, if CD=AB, and the angle to A and C is a right angle, that the angle to B and D is also a right angle. Whence it is at the same time proved, that BD is also a straight line, since it alone holds itself in the same way to both.

(16) Caeterum ex hoc solo, quod demonstravimus § 13 eductis angulo recto ad AB rectis AC, BD aequalibus, angulos ad C et D rectos esse, sequitur rectas AC, BD non posse concurrere, quia Euclides ostendit propositione 28 primi, si anguli CAB et ABD sint aequales, rectas non posse concurrere. Idem ostendit prop. 27 ejusdem, si recta incidat eodem modo in duas parallelas seu angulos EAC, EBD faciat aequales, rectas AC, BD concurrere non posse, seu ipsius definitione esse parallelas. Et haec pendent ex 16. primi Euclidis, quod in triangulo angulus externus sit interno opposito major, α ipsi β vel γ. Nos potuissemus aliter ita demonstrare: concipiendo LM rectam moveri ad AC angulo eodem quocunque, et extremitate L ferri per AC, extremitate M per BD, tunc si ponamus, AC existente recta, etiam BD esse rectam, sequitur eas non concurrere, nam si concurrerent, velut in H, recta LM eo veniens suo situ λμ, faceret angulos λHA, λHB aequales, partem toti. Porro si una sit recta, etiam alteram esse rectam, ex eo sequitur, quod recta LM durante motu ad ambas, quas describit, se habet eodem modo. Itaque supererat solum demonstranda inversa, seu quod rectae non concurrentes inter se ab eadem recta ad eosdem angulos secentur, quae 29. primi; et reapse redit ad Axioma 13. Ad hoc demonstrandum Proclus assumsit, duas rectas, quae a communi producuntur, ad distantiam a se invicem venire quantacunque; Clavius autem rem aliter demonstrare conatur, assumens rectam, quae super alia recta, manente angulo, fertur, describere rectam; id autem nos ex eo demonstramus in hac ipsa paragrapho, quod recta mobilis ad ambas, quas extremis percurrit, se habet eodem modo; ergo si una est recta, etiam altera talis erit.

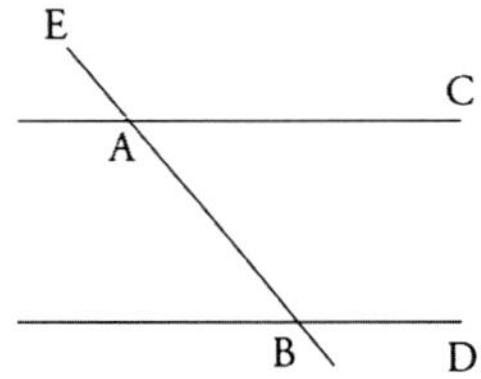

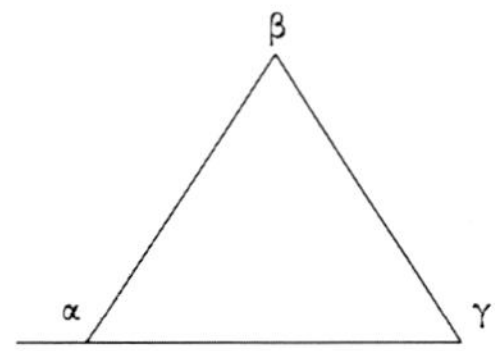

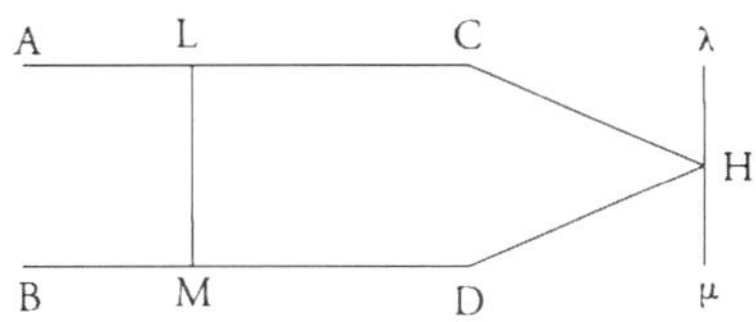

(17) Ut concludatur, possis parallelas definire vel cum Euclide rectas, quae in eodem plano non concurrunt, vel rectas, quae in eandem rectam sunt perpendiculares. Ostendi enim potest, tum has non concurrere, tum non concurrentes esse ad eandem rectam perpendiculares. Ostendi etiam potest, rectas binas, quae ad una aliquam rectam angulos aequales faciunt, ad quemlibet aliam cui occurrunt angulos aequales facere. <Euclidis,> quod parallelae sunt quae non concurrunt; etsi minus sit causalis, habet tamen hoc quod nihil eligit. At quae definit parallelas per eas, quae ad eandem rectam sunt normales, eligit angulum rectum; quae vero vult generatum ut ad eandem rectam faciat eundem angulum, paradoxa est, seu essentiae dubitabilis, quaeritur enim an aliam quemvis alio angulo quo <...>. Si parallelas quis definiat lineas quae nec inter se discerni possunt nec in variis locis respondentibus invicem possint discerni, demonstrare debebit, tales esse rectas.

...

(16) On the other hand, from the fact (which we demonstrated in § 13) that, when the equal straight lines AC and BD are drawn at right angles to AB, also the angles to C and D are right angles, it follows that the straight lines AC, BD cannot come together, since Euclid shows in *Elements* I, 28, if angles CAB and ABD are equal, the straight lines cannot come together. In the same way he shows in *Elements* I, 27, that if a straight line intersects two parallels in the same way, that is, the angles EAC and EBD are equal, then the straight lines AC, BD cannot come together, that is they are parallels following his definition. And this depends on *Elements* I, 16, that in a triangle the external angle is greater than the internal opposite angle, α to β or γ. We can demonstrate it in another way: conceiving LM as a straight line moved with whatever constant angle along AC, and with the extremity L carried through AC, and the extremity M through BD, then if we assume that if AC is a straight line then BD is also a straight line, then it follows that these cannot come together, for if they came together, say in H, the straight line LM coming in the situation $\lambda\mu$, would make angles λHA, λHB equal, the part with the whole. Moreover, the fact that if one is a straight line, the other is also a straight line, follows because the straight line LM, during its motion, holds itself in the same way to both lines which it describes. Thus it remains only to demonstrate the inverse proposition, or that straight lines that don't come together among themselves are cut by the same straight line to the same angles, which is *Elements* I, 29; and in fact we are back to Axiom 13. In order to demonstrate this, Proclus assumed that two straight lines, which are produced from a common point, will come to any distance, no matter how large, from each other; Clavius however tried to demonstrate the matter in another manner, assuming that a straight line, which is carried over another straight line with a constant angle, describes a straight line; we however in this section itself demonstrate this fact, because a movable straight line holds itself in the same way to both straight lines, which it goes through with its extremities; therefore if one of them is a straight line, the other will also be such.

(17) To conclude, you can define parallels either with Euclid as straight lines, which do not come together in the same plane, or as straight lines, which are perpendicular to the same straight line. It can then be shown, in the one case that these do not come together, or in the other that lines which do not come together are perpendicular to the same straight line. It can also be shown, that two straight lines, which make equal angles to one particular straight line, make equal angles to whichever other one that they meet. … For Euclid, that parallels are those which do not come together; although this is less causal, it has however an advantage, that it chooses nothing. But anyone who defines parallels by saying that they are normal to the same straight line, chooses a right angle. And anyone who wants the line generated so that it makes the same angle to the same straight line, takes it as paradoxical notion, or doubtable in essence, since it may be asked whether to any other straight line with another angle which (…). If someone defines parallels as straight lines which cannot be discerned among themselves nor could be discerned the one from the other while regarded from corresponding places, they will need to show that such lines are straight.

…

Ax. XIII.

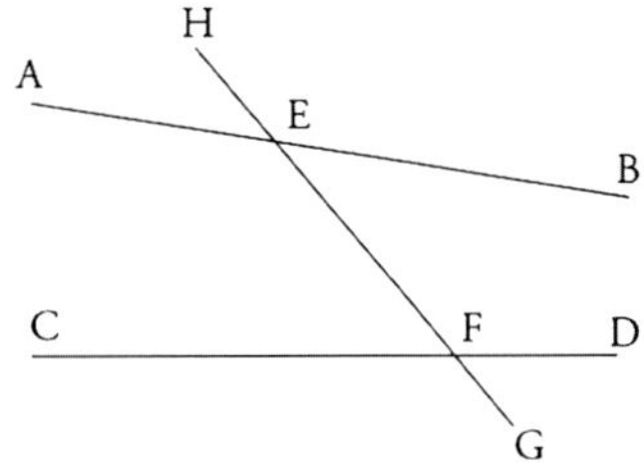

Si in duas rectas lineas altera recta incidens, internos ad easdemque partes angulos duobus rectis minores faciat, duae illae rectae lineae in infinitum productae sibi mutuo incident ad eas partes, ubi sunt anguli duobus rectis minores.

(1) Nempe si anguli BEF et DFE simul sumti sint duobus rectis minores, concurrent rectae AEB, CFD versus B sive D. Hoc pronunciatum potuisset enuntiari clarius hoc modo: Si duae rectae ad eandem non faciant angulos aequales, concurrent. Nam quod concurrant ab ea parte, ubi summa eorum angolorum internorum seu se respicientium est minor, seu ubi ad se inclinantur, facilius demonstrari poterat, nec credo axioma addi opus habebat. Producatur EF ad G et FE ad <H>; si jam anguli BEF, DFG sunt inaequales, erunt BEF+DFE minores duobus rectis. Si BEF−DFG=0, hinc cum sit DFG=2 rect.−DFE, sint BEF−2 rect.+DFE=0; ergo cum recta EF aequales facit angulos ad easdem partes ad AB et CD, erunt BEF+DFE interni aequales duobus rectis; sin non facit aequales, etiam haec summa duobus rectis inaequalis erit, ab una parte minor, ab altera major. Ergo si verum est in casu summae duobus rectis inaequalis concursum etiam in casu rectae ad duos rectos angulos inaequales facientes, earum duarum rectarum concursu erit.

(2) Hujus axiomatis demonstrationem dedit Proclus, credo et ante ipsum Geminus; aliam Clavius tentavit. Sed de his agemus, et breviora etiam tentabimus suo loco.

Ax. XIV.

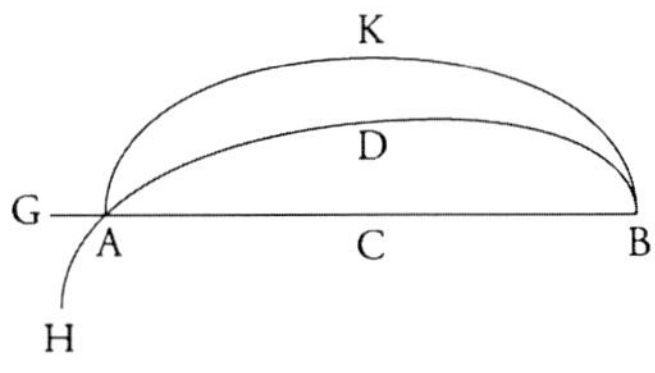

Duae rectae lineae spatium non comprehendunt.

... (5) Caeterum haec etiam ex principio rationis determinantis demonstrare licebit. Nam posito GACB el KADB, GA et HA rursus concurrere in B, non potest dari ratio definiens, quanta sit distantia per puncta A et B. Dicat aliquis eam crescere angulo GAH vel DAC; sed oportet dari legem relationis, utrum nempe crescat in ratione angulorum, an ut quadrata eorun vel cubi vel in alia relatione quacunque, seu quaenam sit linea, in qua anguli poni possunt ut abscissae, et ipsae distantiae ut ordinatae. Praeterea si angulis variatis varietur CB, non posset tertia recta AKB concurrere ipsi ACB in B, quia angulus KAC major eat angulo DAC, et tamen talis dari debet recta AKB, ut supra ostendimus, nam ipsa ADB planum secat bifariam, itaque debet dari AKB, quae eodem modo se habeat ad ADB, ut ACB se ad ADB habet. Cum ergo distantia puncti B a puncto A ex angulo rectarum CA et DA definiri non possit, qui tamen determinat rectarum CA et DA situm ad invicem, atque adeo nulla est ratio determinandi diatantiam punctorum duplicis concursus.

Axiom 13.

If a straight line intersecting with two straight lines, makes on the same side internal angles less than two right angles, when these two lines are produced infinitely they will meet each other mutually on those sides, where the angles are less than two right angles.

(1) Namely, if angles BEF and DFE are together less than two right angles, the straight lines AEB and CFD come together towards B or D. This assertion could have been put more clearly in this way: if two straight lines do not make equal angles to the same straight line, they come together. For that they come together on that side, where the sum of their internal angles is less, or where they incline towards each other, can easily be demonstrated, and I do not believe it necessary to add this to the axiom. Let EF be produced towards G and FE towards H; if indeed angles BEF, DFG are unequal, BEF + DFE will be less than two right angles. If BEF minus DFG equals 0, hence since DFG equals two right angles minus DFE, then BEF minus two right angles plus DFE equals 0; therefore since the straight line EF makes equal angles to the same sides to AB and CD, then the internal angles BEF plus DFE were equal to two right angles. If they do not make equal angles, then this sum will also be unequal to two right angles, on one side less, on the other more. Therefore if indeed the intersection happens in the case of a sum unequal to two right angles, then also in the case in which the straight lines make angles unequal to two right angles, the intersection of these two straight lines will happen.

(2) Proclus gave a demonstration of this axiom, and I believe also Geminus before him; Clavius tried another one. But we will talk about this, and we will indeed try to do it more briefly, in its own place.

...

Axiom 14.

Two straight lines do not enclose space.

... (5) On the other hand, this may also be proved from the principle of determinant reason. Indeed, if we assume that in GACB and HADB, the straight lines GA and HA meet again in B, no reason can be given determining the distance between A and B. Someone might say that the distance will grow with the angle GAH or DAC. Yet, it would be necessary to give a law for this relation, e. g. whether the distance grows proportionally with the angles, or the square of them, or the cube of them, or any other relation expressing which is the line, in which the angles may be assumed as the abscissae and the distances as the ordinates. Moreover, if a variation of the angle would entail a variation of CB, a third straight line AKB could not be given, that meets ACB in the same B, since the angle KAC is greater than the angle DAC. Nonetheless, such a straight line AKB could be given, as we proved above, since the straight line ADB cuts the plane in two identical parts and thus a straight line AKB should be given, that holds itself to ADB in the same way as ACB holds to ADB. Since then the distance between the point A and the point B cannot be defined by the angle formed by the straight lines CA and DA, which however determines the reciprocal situation of the straight lines CA and DA, there is no reason determining the distance between the points of the double meeting.

6.37 De Risi 2007, p. 611.

Recta est linea per puncta sui ad duo proposita puncta situs unica. Recta quaevis terminata est minima linea interdum extrema.

Duae rectae sunt similes inter se.

Si duae rectae se ubique se habeant inter se eodem modo parallelae appellantur, itaque ubique sunt aequidistantes, et duae aliae inter se parallelae se eodem modo habent ad ambas.

6.38 De Risi 2007, p. 613.

Cum duae rectae ubique eodem modo se invicem habent sunt parallelae, id est: si puncto uno rectae A aliquo modo se habente ad rectam B, quodlibet punctum rectae A eodem modo se habebit ad B, recta A erit parallela ad rectam B. Esto $Y.\bar{Z} \simeq B.\bar{Z}$ et sit $\bar{Z}$ recta, erit $\bar{Y}$ recta parallela ipsi $\bar{Z}$. Eadem definitio etiam in planis parallelis et in circumferentiis parallelis (seu concentricis) et superficiebus sphaericis parallelis (seu concentricis) locum habet.

Si sit $A.\bar{Z} \simeq B.\bar{Z}$ et sit $\bar{Y}$ locus unicorum ad A.B, erit $\bar{Y}$ parallela ipsi $\bar{Z}$ et vicissim. Sint L, M, puncta in $\bar{Z}$, sint $A.L.M \simeq B.L.M$, $A.L \simeq B.L$ et $A.M \simeq B.M$. Ergo componendo aliter $A.B.L \simeq B.A.M$. Ergo $\bar{Y}.L \simeq \bar{Y}.M$.

Memorabilis haec consequentia $A.L.M \simeq B.L.M$ et $A.B.L \simeq B.A.M$ aequivalent: resolvendo scilicet alterutrum in aequirelationes partiales, et aliter recomponendo.

6.39 De Risi 2007, p. 618.

Describemus circulum in plano, seu lineam in plano descriptam eodem modo ad unum punctum plani se habentem. Sumtis tribus punctis in circulo et ducta recta inter duo puncta media, et inter alia puncta etiam media, hae rectae concurrent in centro. (Adeoque non possunt esse parallelae, secus foret, si tres rectae fuissent sumtae in recta, seu in circulo infinitè longinquum centrum habente, quo casu fierent parallelae. NB.) Hinc tribus punctis datis determinatus est circulus.

6.37 From *Spatium absolutum* … (August–September 1714)

A straight line is a line which is unique for situation through its own points to two proposed points. Any bounded straight line whatsoever is the shortest line between the extremes.

Two straight lines are similar to one another.

If two straight lines hold themselves everywhere in the same way to one another, they are called parallels, and thus they are everywhere equidistant, and two other lines parallel to one another hold themselves in the same way to both.

6.38 From *Rectam definio* … (1715)

When two straight lines hold themselves everywhere in the same way to one another, they are parallel, that is: if, whenever one point of a straight line A holds itself in some way to a straight line B, any other point of straight line A will hold itself in the same way to B, then the straight line A will be parallel to straight line B. Let $Y.\bar{Z} \simeq B.\bar{Z}$ and let $\bar{Z}$ be a straight line, then $\bar{Y}$ will be parallel to $\bar{Z}$. The same definition has an application also in parallel planes and in parallel (or concentric) circumferences and in parallel (or concentric) spherical surfaces.

If $A.\bar{Z} \simeq B.\bar{Z}$ and if $\bar{Y}$ is the locus of the points unique to A.B, then $\bar{Y}$ will be parallel to $\bar{Z}$ and conversely. If L, M are points in $\bar{Z}$, and $A.L.M \simeq B.L.M$, then $A.L \simeq B.L$ and $A.M \simeq B.M$. Therefore by compounding differently $A.B.L \simeq B.A.M$. Therefore $\bar{Y}.L \simeq \bar{Y}.M$. This consequence is memorable, that $A.L.M \simeq B.L.M$ and $A.B.L \simeq B.A.M$ are equivalent: by resolving indeed one of the two in partial equirelations, and compounding again in another way.

6.39 From *Calculi situs fundamenta* … (1715)

We describe a circle in a plane, or a line described in a plane holding itself in the same way to one point of the plane. When three points are taken in a circle, and a straight line is drawn in the middle between two points, and another line in the middle between the other points, then these straight lines come together in the center. (And therefore these cannot be parallel. It would be otherwise, if the three straight lines were taken in a straight line, or in a circle which had an infinitely distant center, in which case they would be parallel. NB.) Hence a circle is determined from three given points.

Bibliography

A G.W. LEIBNIZ, *Sämliche Schriften und Briefe*, Darmstadt/Leipzig/Berlin, Akademie 1923-.
C G.W. LEIBNIZ, *Opuscules et fragments inédits*, ed. L. Couturat, Paris, Alcan 1903.
GM G.W. LEIBNIZ, *Mathematische Schriften*, ed. C.I. Gerhardt, Berlin/Halle 1849–1863.
GP G.W. LEIBNIZ, *Die philosophischen Schriften*, ed. C.I. Gerhardt, Berlin, Weidmann 1875–1890.
KGS I. KANT, *Gesammelte Schriften*, Berlin/Leipzig, deGruyter 1900-.

F. ACERBI, *Homeomeric lines in Greek mathematics*, "Science in Context", 23 (2010), pp. 1–37.

J. d'ALEMBERT, *Mélanges de literature, d'histoire et de philosophie*, vol. 5, Amsterdam, Chatelain 1767.

A. ARANA, P. MANCOSU, *Descartes and the cylindrical helix*, "Historia Mathematica", 37 (2010), pp. 403–427.

A. ARNAULD, P. NICOLE, *La logique ou l'art de penser*, Paris, Desprez 1662, 16832; reprint by P.Clair, F. Girbal, Paris, PUF 1965.

A. ARNAULD, *Nouveaux elémens de géometrie*, Paris, Savreux 16671, 16832 ; reprint in *Géométries de Port-Royal*, ed. D. Descotes, Paris, Champion 2009.

R.T.W. ARTHUR, *Infinite Aggregates and Phenomenal Wholes: Leibniz's Theory of Substance as a Solution to the Continuum Problem*, "Leibniz Society Review", 8 (1998), pp. 25–45.

R.T.W. ARTHUR, *Infinite Numbers and the World Soul; in Defence of Carlin and Leibniz*, "The Leibniz Review", 9 (1999), pp. 105–16.

R.T.W. ARTHUR, *Leibniz on Infinite Numbers, Infinite Wholes, and the Whole World: A Reply to Gregory Brown*, "The Leibniz Review", 11 (2001), pp. 103–16. [A]

R.T.W. ARTHUR, *Leibniz: The Labyrinth of the Continuum*, New York, Yale University Press 2001. [B]

R.T.W. ARTHUR, *Leibniz's syncategorematic infinitesimals*, "Archive for History of Exact Sciences", 67 (2013), pp. 533–93. [A]

R.T.W. ARTHUR, *Leibniz's Theory of Space*, "Foundations of Science" 18 (2013), pp. 499–528. [B]

U. BALDINI, P.D. NAPOLITANI (ED.),*Christoph Clavius. Corrispondenza*, Pisa, Department of Mathematics (preprint) 1992.

W.H. BARBER, *Leibniz in France, from Arnauld to Voltaire. A Study in French Reactions to Leibnizianism, 1670–1760*, Oxford, OUP 1955.

I. BARROW, *Lectiones Mathematicae XXIII in quibus Principia Matheseos generalia exponuntur*, London, Playford 1683; reprint in BARROW, *The Mathematical Works*, ed. W. Whewell, Cambridge, CUP 1860.

P. BEELEY, S. PROBST, *John Wallis (1616–1703): Mathematician and Divine*, in *Mathematics and the Divine*, eds. L. Bergmans, T. Koetsier, Amsterdam, Elsevier 2005, pp. 441–57.

D. BEESON, *Maupertuis: an intellectual biography*, Oxford, OUP 1992.

F.D. BEHN, *Dissertatio mathematica sistens linearum parallelarum proprietates*, Jena, Margraf 1761.

V. De Risi, *Leibniz on the Parallel Postulate and the Foundations of Geometry*,
Science Networks. Historical Studies, DOI 10.1007/978-3-319-19863-7

Y. Belaval, *Note Sur la pluralité des espaces possibles d'après la philosophic de Leibniz*, "Perspektiven der Philosophie", 4 (1978), pp. 9–19.
J. Bernoulli, *Virorum celeberr. Got. Gul. Leibnitii et Johan. Bernoulli Commercium philosophicum et mathematicum*, Lausanne, Bousquet 1745.
L. Bertrand, *Developpement nouveau de la partie elementaire des mathématiques*, Genève, 1778.
R.O. Besthorn, J.L. Heiberg, *Codex Leidensis 399,1: Euclidis elementa ex interpretatione Al-Hadschdschadschii cum commentariis Al-Nairizii*, Copenhagen, Gyldendel 1893–1932.
G.D. Birkhoff, *The Principle of Sufficient Reason*, "The Rice Institute Pamphlet", 28 (1940), pp. 24–50.
E. Bodemann, *Der Briefwechsel des Gottfried Wilhelm Leibniz in der Königlichen Öffentlichen Bibliothek zu Hannover*, Hannover, Hahn 1889.
F. Bolyai, *Kurzer Grundriss eines Versuchs*, Târgu Mureş, 1851.
B. Bolzano, *Paradoxien des Unendlichen*, ed. F. Prihonsky, Leipzig, Reclam 1851.
R. Bonola, *Un teorema di Giordano Vitale da Bitonto sulle rette equidistanti*, "Bollettino di Bibliografia e Storia delle Scienze Matematiche", 8 (1905), pp. 33–36.
R. Bonola, *La geometria non-euclidea. Esposizione storico-critica del suo sviluppo*, Bologna, Zanichelli 1906. English transl. by H.S. Carslaw, *Non-Euclidean Geometry*, Chicago, Open Court 1912.
K. Bopp, *Antoine Arnauld, der grosse Arnauld, als Mathematiker*, Leipzig 1902.
K. Bopp, *Drei Untersuchungen zur Geschichte der Mathematik*, Berlin, DeGruyter 1929.
G.A. Borelli, *Euclides restitutus, sive priscae Geometriae Elementa brevius et facilius contexta, in quibus praecipue proportionum theoriae nova firmiorique methodo proponantur*, Pisa, Francesco Onofri 16581; Roma, Mascardi 16793. Italian translation by D. Magni (in a 2nd edition, revised by Borelli) as *Euclide rinnovato*, Bologna, Giovan Battista Ferroni 1663.
M.T. Borgato, L. Pepe, *Una memoria inedita di Lagrange sulla teoria delle parallele*, "Bollettino di Storia delle Scienze Matematiche", 8 (1988), pp. 307–335.
J.H.M. Bos, *Differentials, High-Order Differentials and the Derivative in the Leibnizian Calculus*, "Archive for History of Exact Sciences", 14, 1975, pp. 1–90.
F.C.M. Bosinelli, *Proof of Hilbert's Geometrical Axioms of Group I employing Leibniz's Calculus of Situation*, in *Leibniz und Europa. VI. Internationaler Leibniz-Kongreß*, Hannover 1994, pp. 77–84.
G.L.L. de Buffon, *Oeuvres Complètes*, ed. B.G.E. de Lacépède, Paris 1818.
S. Carvallo (ed.), *Stahl-Leibniz. Controverse sur la vie, l'organisme et le mixte*, Paris, Vrin 2004.
U. Cassina, *Sulla dimostrazione di Wallis del Postulato Quinto di Euclide*, in *Actes du 8e Congrès International d'Histoire des Sciences* (1956), Paris, Hermann&Cie 1958, pp. 33–8.
J. Cassinet (ed.), "Cahiers d'histoire des mathématiques de Toulouse", 9 (1986). [A special issue on the transmission to the West of Nasīr ad-Dīn's works.]
E. Cassirer, *Gesammelte Werke*, ed. B. Recki, Hamburg, Meiner 1998–2008.
C. Clavius, *Euclidis Elementorum Libri XV*, Roma, Vincenzo Accolto 15741; Roma, Bartolomeo Grassi 15892; finally in C. Clavius, *Opera mathematica*, Mainz, Reinhard Eltz (et al.) 1611–1612, vol. 1; reprint by E. Knobloch, Hildesheim, Olms 1999.
F. Commandino, *Euclidis Elementorum libri XV, unà cum Scolijs antiquis*, Pesaro, Camillo Franceschini 1572. Italian ed., *Degli Elementi d'Euclide libri quindici*, Urbino, Frisolino 1575.
L. Couturat, *La logique de Leibniz. D'aprés des documents inédits*, Paris, Alcan 1901.
G. Crapulli, *Mathesis Universalis.Genesi di un'idea nel XVI secolo*, Roma, Edizioni dell'Ateneo 1969.
A.L. Crelle, *Ueber Parallelen-Theorien*, Berlin, Maurer 1816.
D.R. Cunningam, C. Mellizo, *Francisco Sánchez: Carta a Cristobal Clavio* "Cuadernos Salmantinos de Filosofía", 5 (1978), pp. 387–406.
M. Curtze, *Anaritii in decem libros priores elementorum Euclidis commentarii ex interpretatione Gherardi Cremonensis*, Leipzig, Teubner 1899.
C.F.M. Dechales, *Cursus seu Mundus Mathematicus*, Lyon, Posuel&Rigaud 1690 (16741).
M. Dehn, *Die Legendre'schen Sätze über die Winkelsumme im Dreieck*, "Mathematische Annalen", 53 (1900), pp. 405–39.

R. Delassus, *De officio medici, sive de vita clarissimi viri Domini Francisci Sanchez*, in F. Sánchez, *Opera medica*, Toulouse, Bosc 1636.

J. Delboeuf, *Prolégomènes philosophiques de la géométrie et solution des postulates*, Liège, Desoer 1860.

V. De Risi, *Leibniz on Geometry. Two unpublished manuscripts with translation and commentary*, "The Leibniz Review", 15, 2005, pp. 127–51.

V. De Risi, *Geometry and Monadology. Leibniz's Analysis Situs and Philosophy of Space*, Basel, Birkhäuser 2007.

V. De Risi, *La dimostrazione kantiana del Quinto Postulato*, in *Kant und die Philosophie in weltbürgerlicher Absicht*, eds. S. Bacin, A. Ferrarin, C. La Rocca, M. Ruffing, Berlin, DeGruyter 2013, vol. 5, pp. 31–43.

V. De Risi (ed.), *Gerolamo Saccheri. Euclid Vindicated from Every Blemish*, Basel, Birkhäuser 2014. [a]

V. De Risi, *Francesco Patrizi e la nuova geometria dello spazio*, in *Locus-Spatium*, eds. D. Giovannozzi and M. Veneziani, Firenze, Olschki, 2014, pp. 269–327. [b]

V. De Risi, *The Development of Euclidean Axiomatics. The systems of principles and the foundations of mathematics in editions of the Elements in the Early Modern Age*, "Archive for History of Exact Sciences", forthcoming. [c]

C.L. Dodgson, *Euclid and his Modern Rivals*, London, MacMillan 1879.

I.E. Drabkin, *Aristotle's wheel: notes on the History of a Paradox*, "Osiris", 9 (1950), pp. 162–98.

J.A. Eberhard, *Ist die Mathematik durch ihre synthetischen Urtheile in Ansehung ihres Wahrheitsgrundes von der Metaphysik verschieden?*, "Philosophisches Magazin", 3.1 (1790), pp. 89–110.

J. Echeverría, *L'Analyse Géométrique de Grassmann et ses rapports avec la Caractéristique Géométrique de Leibniz*, "Studia Leibnitiana", 11 (1979), pp. 223–73.

J. Echeverría, *La caractéristique géométrique de Leibniz en 1679*, Paris, Sorbonne 1980. [PhD diss.]

J. Echeverría, *Investigaciones de Leibniz sobre la nocion y el postulado de las parallelas*, in *Actas del III Congreso de la Sociedad Española de Historia de las Ciencias*, eds. J. Echeverría, M. de Mora Charles, San Sebastián, 1986, vol. 1, pp. 253–63.

J. Echeverría, *La demostracion de los axiomas de Euclides segun Leibniz*, "Revista Latinoamericana de Filosofia", 18, 1992, pp. 33–41.

J. Echeverría, *Leibniz critique d'Euclide: la démonstration des axiomes d'Euclide chez Leibniz*, in *Leibniz und Europa. VI. Internationaler Leibniz-Kongreß*, Hannover 1994, pp. 204–10.

J. Echeverría (ed.), *G.W. Leibniz.La caractéristique géométrique*, Paris, Vrin 1995.

F. Engel, P. Stäckel, *Die Theorie der Parallellinien von Euklid bis auf Gauss*, Leipzig, Teubner 1895.

L. Euler, *Opera omnia*, Leipzig, Teubner 1911-.

H. Fabri, *Synopsis Geometrica*, Lyon, Molin 1669.

M. Fichant, *Leibniz y la exigencia de demostracion de los axiomas*, "Revista Latinoamericana de Filosofia", 18, 1992, pp. 43–82. French version as *Leibniz et l'exigence de démonstration des axiomes : 'La partie est plus petite que le Tout'*, in M. Fichant, *Science et métaphysique dans Descartes et Leibniz*, Paris, PUF, 1998, pp. 329–73.

M. Fichant, *Science et métaphysique dans Descartes et Leibniz*, Paris, PUF, 1998, pp. 329–73.

J.V. Field, J.J. Gray, *The Geometrical Work of Girard Desargues*, Berlin, Springer 1987.

J.V. Field, *The Invention of Infinity: Mathematics and Art in the Renaissance*, Oxford, OUP 1997.

M. Folkerts, *Mittelalterliche mathematische Handschriften in westlichen Sprachen in der Herzog August Bibliothek Wolfenbüttel. Ein vorläufiges Verzeichnis*, "Centaurus", 25 (1981), pp. 1–49.

W.B. Frankland, *Theories of Parallelism*, Cambridge, CUP 1910.

G. Frege, *Die Grundlagen der Arithmetik*, Breslau, Köbner 1884.

G. Freudenthal, *Definition and Construction. Salomon Maimon's Philosophy of Geometry*, Max-Planck-Institut für Wissenschaftsgeschichte, Preprint 317, 2006.

H. Freudenthal, *Leibniz und die Analysis Situs*, "Studia Leibnitiana", 4, 1972, pp. 61–69.

M.N. Fried, S. Unguru, *Apollonius of Perga's Conica: text, context, subtext*, Leiden, Brill 2001.

M. Futch, *Leibniz's metaphysics of Time and Space*, Berlin, Springer 2008.

G. Galilei, *Le opere*, ed. A Favaro, Firenze, Barbera 1890–1909.

D. Garber, *Leibniz and Fardella: Body, Substance and Idealism*, in *Leibniz and his Correspondents*, ed. P. Lodge, Cambridge, CUP 2004, pp. 123–40.

J.-L. Gardies, *Pascal et l'axiome d'Archimède*, "Revue philosophique de la France et de l'Étranger", 171 (1981), pp. 425–40.

C.F. Gauss, *Werke*, Hildesheim, Olms 1975–1987.

V. Giordano, *Euclide Restituto, ovvero gli antichi elementi geometrici ristaurati e facilitati*, Roma, Angelo Bernabò 16801, 16862.

G. Giovannozzi, *La versione borelliana di Apollonio*, "Accademia Romana dei Nuovi Lincei", 2 (1916), pp. 1–31.

E. Giusti, *La géométrie du meilleur des mondes possibles: Leibniz critique d'Euclide*, "Studia Leibnitiana Sonderheft", 21, 1992, pp. 215–32.

E. Giusti, *Euclides Reformatus. La teoria delle proporzioni nella scuola galileiana*, Torino, Bollati Boringhieri 1993.

U. Goldenbaum, *Appell an das Publikum: die öffentliche Debatte in der deutschen Aufklärung 1687–1796*, Berlin, Akademie 2004.

J.V. Grabiner, *Why Did Lagrange "Prove" the Parallel Postulate?*, "The American Mathematical Monthly", 116 (2009), p. 3–18.

H. Grassmann, *Gesammelte mathematische und physikalische Werke*, ed. F. Engel, Leipzig, Teubner 1896–1911.

J. Gray, L. Tilling, *Johann Heinrich Lambert, mathematician and scientist, 1728–1777*, "Historia Mathematica", 5 (1978), pp. 13–41.

M.J. Greenberg, *Euclidean and Non-Euclidean Geometries without Continuity*, "The American Mathematical Monthly", 86 (1979), pp. 757–64.

M.J. Greenberg, *Aristotle's Axiom in the Foundations of Geometry*, "Journal of Geometry", 33 (1988), pp. 53–57.

M.J. Greenberg, *Old and New Results in the Foundations of Elementary Plane Euclidean and Non-Euclidean Geometries*, "The American Mathematical Monthly", 117 (2010), pp. 198–219.

L. Guerrini, *Matematica ed erudizione. Giovanni Alfonso Borelli e l'edizione fiorentina dei libri V, VI e VII delle Coniche di Apollonio di Perga*, "Nuncius", 14 (1999), pp. 505–568.

G.F. Hagen, *Mediationes philosophicae de methodo mathematica*, Nürnberg, Stein 1734.

M. Hallett, U. Majer (eds.), *David Hilbert's Lectures on the Foundations of Geometry 1891–1902*, Berlin, Springer 2004.

O. Harari, *Proclus' Account of Explanatory Demonstrations in Mathematics and its Context*, "Archiv für Geschichte der Philosophie", 90 (2008), pp. 137–164.

R. Hartshorne, *Geometry: Euclid and Beyond*, New York, Springer 2000.

C.A. Hausen, *Elementa matheseos*, Leipzig, Marche 1734.

T. Hayashi, *Introducing Movement into Geometry: Roberval's influence on Leibniz's Analysis Situs*, "Historia Scientiarum", 8 (1998), pp. 53–69.

A.E. Heath, *The Geometrical Analysis of Grassmann and its connection with Leibniz's Characteristic*, "The Monist", 27 (1917), pp. 36–52.

T.L. Heath (ed.), *Euclid. The Thirteen Books of the Elements*, Cambridge, CUP 19252.

A. Heinekamp, *Louis Dutens und seine Ausgabe der Opera omnia von Leibniz*, "Studia Leibnitiana Supplementa", 26 (1986), pp. 1–28.

J. Heis, *Leibniz versus Kant on Euclid's Axiom of Parallels*, forthcoming. [a]

J. Heis, *Kant on Parallel Lines*, forthcoming. [b]

C. Herlinus, K. Dasypodius, *Analyseis geometricae sex librorum Euclidis*, Strasbourg, Rihel 1566.

C.W. Hessling, *Theorie der Parallellinien*, Halle, 1818.

D. Hilbert, *Grundlagen der Geometrie*, Leipzig, Teubner 1899; with revisions and supplementary material by P. Bernays, Stuttgart 1962.

C.F. Hindenburg, *Ueber die Schwürigkeit bey der Lehre von den Parallellinien. Neues System der Parallellinien*, "Leipziger Magazin zur Naturkunde, Mathematik und Oekonomie", 1781, pp. 145–68.

C.F. Hindenburg, *Anmerkungen über das neue System der Parallellinien*, "Leipziger Magazin zur Naturkunde, Mathematik und Oekonomie", 1781, pp. 342–71.

C.F. Hindenburg, *Noch etwas über die Parallellinien*, "Leipziger Magazin für reine und angewandte Mathematik", 1786, pp. 359–404.

T. Hobbes, *Opera philosophica quae latine scripsit omnia*, ed. W. Molesworth, London, Bohn&Longman 1839–1845.

T. Hobbes, *The English Works*, ed. W. Molesworth, London, Bohn&Longman 1839–1845.

J.E. Hofmann, *Leibniz in Paris 1672–1676: his growth to mathematical maturity*, London, CUP 1974.

G. Hon, B.R. Goldstein, *From Summetria to Symmetry: The Making of a Revolutionary Scientific Concept*, New York, Springer 2008.

C. Huygens, *Horologium oscillatorium, sive de motu penduloum ad horologia aptato demonstrationes geometricae*, Paris, Muguet 1673. (Now also in vols. 18–19 of the *Oeuvres complètes*.)

C. Huygens, *Oeuvres complètes*, Den Haag, Nijhoff 1888–1970.

J. Iriarte, *Francisco Sánchez, el escéptico disfrazado de Carnéades, en discussion epistolar con Cristobal Clavio*, "Gregoriana", 21 (1940), pp. 413–451.

A. Jacobi, *De undecimo Euclidis axiomate*, Jena, Crök 1824.

K. Jaouiche, *La théorie des parallèles en pays d'Islam*, Paris, Vrin 1986.

D.M. Jesseph, *Squaring the Circle. The War between Hobbes and Wallis*, Chicago, UCP 1999.

N. Jolley, *Leibniz. Ad Cristophori Stegmanni Metaphysicam Unitariorum*, "Studia Leibnitiana", 7 (1975), pp. 161–89.

V. Jullien, *Les étendues géométriques et la ligne droite de Roberval*, "Revue d'histoire de Sciences", 46 (1993), pp. 493–521

V. Jullien, *Eléments de géométrie de G.P. de Roberval*, Paris, Vrin 1996.

J. Jungius, *Geometria empirica*, Rostock, 1627.

J. Jungius, *Disputatio philosophica de stoecheosi geometrica*, Hamburg, Werner 1634. Now in C. Müller-Glauser, *Jungius. Disputationes Hamburgenses*, Göttingen, Vanderhoeck 1988.

H. Kangro, *Joachim Jungius und Gottfried Wilhelm Leibniz: Ein Beitrag zum geistigen Verhältnis beider Gelehrten*, "Studia Leibnitiana", 1 (1969), pp. 175–207.

I. Kant, *Gesammelte Schriften*, Berlin, DeGruyter 1966-. [KgS]

W.J.G. Karsten, *Praelectiones matheseos theoreticae elementaris*, Rostok, Berger 1758.

W.J.G. Karsten, *Mathesis theoretica elementaris atque sublimior*, Rostok, Röse 1760.

W.J.G. Karsten, *Versuch einer völlig berichtigten Theorie von den Parallellinien*, Halle, 1778.

W.J.G. Karsten, *Mathematische Abhandlungen*, Halle, Renger 1786.

G.A. Kästner, *Ueber den mathematischen Begriff des Raums*, "Philosophisches Magazin", II, 4 (1790), pp. 403–19.

G.A. Kästner, *Ueber die geometrischen Axiome*, "Philosophisches Magazin", II, 4 (1790), pp. 420–30.

G.A. Kästner, *Anfangsgründe der Arithmetik, Geometrie, ebenen und sphärischen Trigonometrie und Perspectiv*, Göttingen, Vandenhock 1800 (17581).

W. Killing, *Einführung in die Grundlagen der Geometrie*, Padeborn, Schöningh 1893–1898.

G.S. Klügel, *Conatuum praecipuorum theoriam parallelarum demonstrandi recensio*, Göttingen, Schultz 1763.

J. Knar, *Ueber die Theorie der Parallellinien*, "Zeitschrift für Physik und Mathematik", 3 (1827), pp. 414–39.

J. Knar, *Berichtigung meiner Ansicht über die Theorie der Parallellinien*, "Zeitschrift für Physik und Mathematik", 4 (1828), pp. 427–36.

H.H. Knecht, *Leibniz et Euclide*, "Studia Leibnitiana", 6 (1974), pp. 131–43.

E. Knobloch (ed.), *Leibniz. De Quadratura arithmetica circuli ellipseos et hyperbolae cujus corollarium est trigonometria sine tabulis*, Göttingen, Vandenhoeck&Ruprecht 1993.

E. Knobloch, *Galileo and Leibniz: Different Approaches to Infinity*, "Archive for History of Exact Sciences", 54 (1999), pp. 87–99.

E. Knobloch, *La connaissance des mathématiques arabes par Clavius*, "Arabic Science and Philosophy", 12 (2002), pp. 257–84. [a]

E. Knobloch, *Leibniz's rigorous foundation of infinitesimal geometry by means of Riemannian sums*, "Synthese", 133 (2002), pp. 59–73. [b]

E. Knobloch, *Le calcul leibnizien dans la correspondance entre Leibniz et Jean Bernoulli*, in *Neuzeitliches Denken*, eds. G. Abel, H.-J. Engfer, C. Hubig, Berlin, DeGruyter 2002, pp. 173–93. [c]

E. Knobloch, *The notion of mathematics. An historico-epistemological approach using Kaspar Schott's Encyclopedia of all mathematical sciences* "Almagest", 2 (2011), pp. 36–59.

E. Knobloch, *Galilei und Leibniz*, Hannover, Wehrhahn 2012.

S. König, *De Universali Principio Aequilibrii & motus*, "Nova Acta Eruditorum", 1751, pp. 125–35 and 162–76.

S. König, *Appel au Publique*, Leiden 1752.

S. König, *Defense de l'Appel au Publique*, Leiden 1753.

S. König, *Elemens de geometrie*, ed. J.J. Blassière, Den Haag, van Os 1762.

J.L. Lagrange, *Oeuvres*, ed. J.-A. Serret, Paris, Gauthier 1867–1882.

J.H. Lambert, *De universaliori calculi idea*, "Nova Acta Eruditorum", November/December 1765, pp. 441–73.

J.H. Lambert, *Anlage zur Architectonic*, Riga, Hartknoch 1771.

J.H. Lambert, *Theorie der Parallellinien*, "Magazin für reine und angewandte Mathematik", 1786, pp. 13–64, 325–58.

J.H. Lambert, *Deutscher gelehrter Briefwechsel*, ed. J. Bernoulli, Berlin 1781–1787.

J. Lauter, *Die Prinzipien und Methoden der Geometrie bei Leibniz*, Aachen 1953. [PhD diss.]

G. Lechalas, *Une définition géométrique du plan et de la ligne droite apres Leibniz et Lobatchewsky*, "Revue de metaphysique et de morale", 20 (1912), pp. 718–21.

A.-M. Legendre, *Réflexions sur différentes manières de démontrer la théorie des parallèles ou le théorème sur la somme des trois angles du triangle*, "Mémoires de l'Académie Royale des Sciences de l'Institut de France", 12 (1833), pp. 367–410.

J.W.H. Lehmann, *Mathematische Abhandlungen betreffend die Begründung und Bearbeitung verschiedener mathematischer Theorieen*, Zerbst, Kummer 1829.

G.W. Leibniz, *Oeuvres philosophiques latines et françaises du feu Mr. L.*, ed. E. Raspe with a preface by A. Kästner, Amsterdam/Leipzig, Schreuder 1765.

G.W. Leibniz, *Opera Omnia*, ed. L. Dutens, Genève, Tournes 1768.

G.W. Leibniz, *Briefwechsel zwischen Leibniz und Ch. Wolff*, ed. C.I. Gerhardt, Halle 1860.

G.W. Leibniz, *Der Briefwechsel von Gottfried Wilhelm Leibniz mit Mathematikern*, ed. C.I. Gerhardt, Berlin, Mayer&Müller 1899.

S. Levey, *Leibniz on Mathematics and the Actually Infinite division of matter*, "The Philosophical Review", 107 (1998), pp. 49–96.

T. Lévy, *Gersonide, le Pseudo-Ṭūsī, et le postulat des parallèles. Les mathématiques en Hébreu et leurs sources arabes*, "Arabic Science and Philosophy", 2 (1992), pp. 39–82.

N.I. Lobachevsky, *Pangeometry*, ed. A. Papadopoulos, Zürich, EMS 2010.

A. Lo Bello (ed.), *The Commentary of Al-Nayrizi on Book I of Euclid's Elements of Geometry*, Leiden, Brill 2003 [a].

A. Lo Bello (ed.), *Gerard of Cremona's Translation of the Commentary of Al-Nayrizi on Book I of Euclid's Elements of Geometry*, Leiden, Brill 2003 [b].

J.L. Loemker (ed.), *Leibniz. Philosophical Papers and Letters*, London, CUP 1956.

B. Look, D. Rutherford (eds.), *The Leibniz – Des Bosses Correspondence*, Yale, YUP 2007.

J.F. Lorenz, *Grundriß der reinen und angewandten Mathematik*, Helmstädt, Fleckeisen 1791.

L. Maierù, *Il quinto postulato euclideo in Cristoforo Clavio*, "Physis", 20 (1978), pp. 191–212.

S. Maimon, *Kritische Untersuchungen über den menschlichen Geist*, Leipzig, Fleischer 1797.

N. Malebranche, *Oeuvres complètes*, Paris, Vrin 1958–1978.

P. Mancosu, *Grundlagen, Section 64: Frege's discussion of definitions by abstraction in historical context*, "History and Philosophy of Logic", 36 (2014), pp. 62–89.

A. Marchetti, *La natura della proporzione e della proporzionalità con nuovo, facile, e sicuro metodo*, Pistoia, Gatti 1695.

A. Marchetti, *Euclides Reformatus*, Livorno, Celsi 1709.

R.B. McClenon, *A Contribution of Leibniz to the History of Complex Numbers*, "The American Mathematical Monthly", 30 (1923), pp. 369–74.
H. Meissner, *Des gantzen, in XV. Büchern bestehenden, teutschen Euclidis erstes Buch*, Hamburg, 1696.
N. Mercator, *Euclidis Elementa Geometrica novo Ordine ac Methodo fere demonstrata*, London, Martyn 1678.
J. Mesnard, *Leibniz et les papiers de Pascal*, "Studia Leibnitiana Supplementa", 17 (1978), pp. 45–58.
D.M. Miller, *Representing Space in the Scientific Revolution*, Cambridge, CUP 2014.
I. Mueller, *Philosophy of Mathematics and Deductive Structure in Euclid's Elements*, Cambridge, MIT Press 1981.
M. Mugnai, *Leibniz's Theory of Relations*, "Studia Leibnitiana Supplementa", 28, 1992.
M. Mugnai, *Logic and Mathematics in the Seventeenth Century*, "History and Philosophy of Logic", 31 (2010), pp. 297–314.
H.P. Münzenmayer, *Der Calculus Situs und die Grundlagen der Geometrie bei Leibniz*, "Studia Leibnitiana", 11, 1979, pp. 274–300.
Nasīr ad-Dīnat-tūsī, *Euclidis elementorum geometricorum libri tredecim ex traditione doctissimi Nasiridini Tusini*, Roma, Typographia Medicea 1594.
I. Newton, *Philosophiae naturalis Principia Mathematica*, 3rd ed. London, Innys 1726; eds. A. Koyré, I.B. Cohen, A. Whitman, Cambridge, CUP 1972.
I. Newton, *De gravitatione et aequipondio fluidorum*, in *Unpublished scientific papaers of Isaac Newton*, eds. A.R Hall, M.B. Hall, Cambridge, CUP 1962, pp. 89–156.
E. Olaso, *Francisco Sanches e Leibniz*, "Análise" 4 (1986), pp. 37–74.
A.R.E. Oliveira, *Lagrange as a Historian of mechanics*, "Advances in Historical Studies", 2 (2013), pp. 126–30.
M. Ohm, *Kritische Beleuchtungen der Mathematik überhaupt und der Euklidischen Geometrie insbesondere*, Berlin, Maurer 1819.
V. Pambuccian, *Axiomatizations of Hyperbolic and Absolute Geometries*, in *Non-Euclidean Geometries*, eds. A. Prékopa, E. Molnár, New York, Springer 2006, pp. 119–53.
B. Pascal, *Oeuvres*, eds. L. Brunschvicg, P. Boutroux, F. Gazier, Paris, Hachette 1904–1914.
F. Patrizi, *Della nuova geometria libri XV*, Ferrara, Baldini 1587.
F. Patrizi, *De rerum natura libri II priores. Alter de spacio physico, alter de spacio mathematico*, Ferrara, Baldini 1587.
F. Patrizi, *Nova de universis philosophia*, Ferrara, Mammarelli 1591.
J. Peletier, *Demonstrationum in Euclidis elementa geometrica libri sex*, Lyon, Tornes 1557.
A. Pelletier, *Les Réflexions mathématiques de Kant (1764–1800)*, "Les cahiers philosophiques de Strasbourg", 26 (2009), pp. 97–116.
J.-C. Pont, *L'aventure des parallèles. Historie de la géométrie non euclidienne: précurseurs et attardés*, Berne, Lang 1986.
H. Pulte, *Das Prinzip der kleinsten Wirkung und die Kraftkonzeptionen der rationalen Mechanik*, "Studia Leibnitiana Sonderheft", 19 (1989).
D. Rabouin, *Mathesis Universalis. L'idée de "mathématique universelle" d'Aristote à Descartes*, Paris, PUF 2009.
D. Rabouin, *The Difficulty of Being Simple: Interactions between Mathematics and Philosophy in Leibniz's Analysis of Notions*, in *G.W. Leibniz, Interrelations between Mathematics and Philosophy*, eds. N. Goethe, P. Beeley, D. Rabouin, New York, Springer 2015.
R. Rashed, C. Houzel, *Thābit ibn Qurra et la théorie des parallèles*, "Arabic Science and Philosophy", 15 (2005), pp. 9–55.
P.-S. Régis, *Cours entier de philosophie ou systeme general selon les principes de M. Descartes*, Lyon, Posuel&Rigaud 1691 (16901).
H. Reichardt, *Gauß und die Anfänge der nicht-euklidischen Geometrie*, Leipzig, Teubner 1985.
P. Remnant, J. Bennett (eds.), *Leibniz. New Essays on Human Understanding*, Cambridge, CUP 1996.
N. Rescher, *Leibniz's Metaphysics of Nature*, Dordrecht, Reidel 1981.

C. Richard, *Euclidis elementorum geometricorum libros tredecim*, Antwerp, Verdus 1645.

W. Risse, *Joachim Jungius. Logicae Hamburgensis additamenta*, Göttingen, Vandenhoeck 1977.

A. Robinet, *Correspondance Leibniz-Clarke*, Paris, PUF 1957.

A. Robinet, *G.W. Leibniz. Phoranomus seu de potentia legibus naturae*, "Physis", 28 (1991), pp. 429–541 and 797–885.

B.A. Rosenfeld, *A History on Non-Euclidean Geometry. Evolution of the Concept of a Geometric Space*, transl. A. Shenitzer, New York, Springer 1988.

B. Russell, *A critical exposition of the philosophy of Leibniz*, Cambridge, CUP 1900.

A.I. Sabra, *Thābit Ibn Qurra on Euclid's Parallels Postulate*, "Journal of the Warburg and Courtauld Institutes", 31 (1968), pp. 12–32.

A.I. Sabra, *Simplicius's Proof of Euclid's Parallels Postulate*, "Journal of the Warburg and Courtauld Institutes", 32 (1969), pp. 1–24.

G. Saccheri, *Logica Demonstrativa*, Torino, Paolino 1697; reprint by W. Risse, Hildesheim, Olms 1980.

G. Saccheri, *Neo-Statica*, Milano, Malatesta 1708.

G. Saccheri, *Euclides ab omni naevo vindicatus*, Milano, Montani 1733.

F. Sánchez, *Opera philosophica*, ed. J. De Carvalho, Coimbra, 1955.

J. Sauveur, *Geometrie élémentaire et pratique*, ed. Le Blonde, Paris, Rollin 1753.

C. Schott, *Cursus mathematicus, sive Absoluta omnium mathematicarum disciplinarum encyclopaedia*, Würzburg, Hertz 1661.

H. Schotten, *Inhalt und Methode des planimetrischen Unterrichts*, Leipzig, Teubner 1890–1893.

G. Schubring, *Die Theorie des Unendlichen bei Johann Schultz (1739–1805)*, "Historia Mathematica", 9 (1982), pp. 441–84.

J.F. Schultz, *Erläuterungen über des Herrn Professor Kant Critik der reinen Vernunft*, Königsberg, Dengel 1784.

J.F. Schultz, *Entdeckte Theorie der Parallelen*, Königsberg, Kanter 1784

J.F. Schultz, *Darstellung der vollkommenen Evidenz und Schärfe seiner Theorie der Parallelen*, Königsberg, Lebrecht 1786.

J.F. Schultz, *Versuch einer genauen Theorie des Unendlichen*, Königsberg, Lebrecht 1788.

J.F. Schultz, *Prüfung der Kantischen Critik der reinen Vernunft*, Königsberg, Hartung 1789–1792.

J.C. Schwab, *Tentamen novae parallelarum theoriae notione situs fundatae*, Stuttgart, Erhard 1801.

J.C. Schwab, *Essai sur la situation, pour servir de supplément aux principes de la géométrie*, Stuttgart, Cotta 1808.

J.C. Schwab, *Commentatio in primum elementorum Euclidis librum*, Stuttgart, Steinkopf 1814.

F.C. Schweikart, *Die Theorie der Parallellinien*, Jena, Gabler 1807.

J.A. Segner, *Vorlesungen über die Rechenkunst und Geometrie*, Lemgo, Meyer 1747.

J.A. Segner, *Cursus mathematici pars I. Elementa arithmeticae, geometriae et calculi geometrici*, Halle, Renger 1756.

J.A. Segner, *Anfangsgründe der Arithmetick, Geometrie, und der geometrischen Berechnungen*, Halle, Renger 1764.

D. Sherry, M. Katz, *Infinitesimals, Imaginaries, Ideals and Fictions*, "Studia Leibnitiana", 44 (2012), pp. 166–92.

B.H. Sude, *Ibn al-Haytham Commentary on the Premises of Euclid's Elements*, Princeton, Princeton University 1974. [PhD Diss.]

R. Taton, *L'oeuvre de Pascal en géométrie projective*, in *L'oeuvre scientifique de Pascal*, Paris, PUF 1964, pp. 17–72.

R. Taton, *L'initiation de Leibniz à la géométrie (1672–1676)*, "Studia Leibnitiana Supplementa", 17 (1978), pp. 45–58.

R. Taton, *Lagrange et Leibniz. De la théorie des fonctions au principe de raison suffisante*, "Studia Leibnitiana Supplementa", 26 (1986), pp. 139–47.

L. Tenca, *Relazioni fra Gerolamo Saccheri e il suo allievo Guido Grandi*, "Studia Ghisleriana", 1 (1952), pp. 19–45.

G. Tonelli, *Leibniz on Innate Ideas and the Early Reactions to the Publication of the Nouveaux Essais (1765)*, "Journal of the History of Philosophy", 12 (1974), pp. 437–54.

P.M.J.E. Tummers, *The Latin Translation of Anaritius' Commentary on Euclid's Elements of Geometry, Books I–IV*, Nijmegen, Ingenium 1994.

C.C.H. Vermehren, *Versuch die Lehre von der parallelen und convergenten Linien*, Güstrow, Ebert 1816.

G. Veronese, *Fondamenti di geometria*, Padova, Tipografia del Seminario 1891.

K. Volkert, *Das Undenkbare denken. Die Rezeption der nichteuklidischen Geometrie im deutschsprachigen Raum (1860–1900)*, Berlin, Springer 2013.

J. Wallis, *De Postulato Quinto et Definitione Quinta Libri 6 Euclidis Disceptatio geometrica*, in Wallis, *Opera mathematica*, Oxford, Theatrum Sheldonianum 1693–1699, vol. 2, pp. 665–78; reprint by C.J. Scriba, Hildesheim, Olms 1972.

J. Wallis, *Elenchus geometriae hobbianae*, London, Hall 1655.

G.G. Wallwitz, *Strukturelle Probleme in Leibniz' Analysis Situs*, "Studia Leibnitiana", 23, 1991, pp. 111–18.

J. Webb, *Immanuel Kant and the Greater Glory of Geometry*, in *Naturalistic Epistemology*, eds. D. Nails, A. Shimony, Dordrecht, Reidel 1987, pp. 17–70.

C. Wolff, *Anfangs-Gründe aller mathematischen Wissenschafften*, Leipzig, Renger 1738 (17101).

C. Wolff, *Elementa Matheseos universae*, Halle, 1713–1741.

C. Wolff, *Vernünfftige Gedancken von den Kräften des menschlichen Verstand*, Halle 1754 (17121).

C. Wolff, *Philosophia rationalis sive Logica*, Leipzig, Renger 1728.

K. Zormbala, *Gauss and the Definition of the Plane Concept in Euclidean Elementary Geometry*, "Historia Mathematica", 23 (1996), pp. 418–36.

Index of Names

A
Acerbi, 94
Aganis, 93
Agapius, 8, 9
Alfonso de Valladolid, 10
al-Haytham, 60
an-Nayrīzī, 8, 9, 93
Apollonius, 9, 25, 26, 76, 82, 159, 167
Arana, 94
Archimedes, 27, 30, 53, 54, 59, 60, 64, 66, 83, 100, 118
Aristotle, 17, 21, 28, 36, 43, 58, 83, 92, 111
Arnauld, 13, 22, 23, 24, 26, 28, 36, 38, 46, 62, 81, 84, 90, 147
Arthur, vi, 42, 67, 68, 127

B
Baldini, 9, 81
Barber, 104
Barrow, 28, 31, 65, 71, 96, 129
Beeley, 19
Beeson, 104
Behn, 108
Belaval, 49
Bennett, 38
Bernoulli, 16, 22, 26, 58, 82, 84, 85, 103, 145, 157, 167
Bertrand, 114
Besthorn, 8
Birkhoff, 52
Bodemann, 16
Bodenhausen, 30
Bolyai, Farkas, 93
Bolyai, János, 50, 106, 120
Bolzano, 115
Bonola, 7, 15, 97, 117, 119
Bopp, 46, 90
Borelli, 14, 15, 16, 18, 21, 29, 30, 34, 38, 39, 47, 51, 58, 59, 67, 75, 76, 78, 79, 82, 97, 98, 107, 110
Borgato, 119, 120
Bos, 64
Boscovich, 109
Bosinelli, 4
Breger, vi
Buffon, 104

C
Caesar, 54
Camus, 109
Cantor, 67
Carroll, 113
Carvallo, 51
Cassina, 64
Cassinet, 9, 17
Cassirer, 4, 41, 46, 72
Cavalieri, 62, 65
Ceva, 17
Cheyne, 51
Clairaut, 109
Clarke, 42, 43, 51, 53, 54, 110
Clavius, 7, 8, 9, 10, 11, 12, 13, 15, 16, 18, 22, 24, 25, 27, 28, 29, 31, 37, 38, 42, 43, 47, 58, 59, 65, 72, 73, 74, 75, 80, 81, 82, 86, 87, 89, 91, 92, 93, 94, 95, 96, 100, 101, 102, 131, 133, 137, 143, 153, 155, 159, 167, 169, 175, 177
Clifford, 95
Commandino, 8, 9, 10, 43, 80, 92
Conring, 22, 33
Couturat, 4, 65
Crapulli, 78
Crelle, 114
Cunningam, 81
Curtze, 93

D
D'Alembert, 57, 70, 97, 120
Dasypodius, 28, 29
Dechales, 12, 71
Dedekind, 44
Dehn, 64
Delassus, 81
Delboeuf, 50
De Risi, 4, 11, 16, 22, 24, 30, 42, 43, 44, 47, 48, 50, 51, 53, 55, 58, 59, 60, 61, 72, 73, 84, 94, 96, 100, 108
Desargues, 62, 69, 70, 110
Des Bosses, 12, 49
Descartes, 25, 36, 42, 65, 94
Descotes, 13, 23, 38, 46
De Volder, 45
Dodgson, 113
Drabkin, 83
Duke of Hannover, 30
Dutens, 120

E
Eberhard, 114
Echeverría, 4, 24, 29, 30, 40, 44, 47, 48, 63, 71, 72, 94
Engel, 116
Euclid, 3, 4, 5, 6, 7, 8, 9, 10, 11, 12, 13, 14, 15, 16, 18, 21, 22, 23, 24, 25, 26, 27, 28, 29, 30, 31, 33, 34, 36, 37, 38, 39, 40, 43, 44, 46, 47, 50, 52, 53, 54, 56, 57, 58, 59, 60, 61, 63, 64, 65, 66, 67, 68, 69, 70, 71, 73, 74, 76, 77, 79, 80, 81, 83, 84, 87, 89, 91, 92, 93, 94, 95, 96, 97, 98, 99, 100, 101, 102, 105, 106, 107, 108, 113, 116, 119, 120, 129, 131, 133, 135, 137, 143, 149, 153, 155, 159, 161, 163, 165, 167, 169, 173, 175
Euler, 104

F
Fabbrizi, vi
Fabri, 12, 58, 66, 71, 125
Fardella, 16
Fichant, 31
Field, 70
Fogel, 16, 79, 80
Folkerts, 8
Foucher, 26, 37, 65
Fourier, 118
Frankland, 5
Frege, 113
Freudenthal, vi, 4, 12
Fried, 82
Futch, 42

G
Galileo, 14, 82, 94
Gallois, 34, 76, 90
Garahan, vi
Garber, vi, 16
Gardies, 64
Gassendi, 42
Gauss, 52, 60, 87, 108, 109, 116, 120
Geminus, 7, 25, 81, 94, 169, 177
Gerardo da Cremona, 93
Gergonne, 114
Gerhardt, 5, 48, 81, 97
Giordano, 15, 16, 58, 59, 60, 87, 93, 97, 98, 108
Giovannozzi, 82
Giusti, 4, 43, 77
Glezer, vi
Goldenbaum, 104
Goldstein, 74
Grabiner, 119
Grandi, 17
Grassmann, 72
Gray, 70, 118
Graziani, vi
Greenberg, 30, 92
Gregory, 85
Grynaeus, 7, 29
Guerrini, 82
Guldin, 133

H
Hagen, 107
Ha-Lévi, 75
Hall, 42
Hallett, 5
Harari, 36
Hartshorne, 30
Hausen, 108
Hayashi, 60
Heath, 7, 72
Heiberg, 8, 54, 66, 118
Heinekamp, 120
Heis, 5, 39, 108
Helmholtz, 52, 121
Herlinus, 28, 29, 42
Hermann, 85

Heron, 9, 10, 37, 59, 60
Hessling, 114
Hilbert, 5, 64, 68
Hindenburg, 110, 111, 112, 114, 115, 116
Hobbes, 18, 19, 21, 34, 35, 36, 38, 41, 43, 58, 62, 65, 86, 94, 167
Hofmann, 65
Holland, 105
Hon, 74
Houzel, 14
Huygens, 62, 71, 78, 82, 84, 103, 117, 118, 119

I
Iriarte, 81

J
Jacobi, 113
Jaouiche, 8, 9, 37
Jesseph, 19
Jolley, 53
Jost, vi
Jullien, 13, 14, 60
Jungius, 28, 29, 79, 80, 87

K
Kangro, 79
Kant, 5, 14, 29, 37, 39, 77, 104, 105, 108, 114, 115, 116, 118
Karsten, 110, 111, 112, 115, 116
Kästner, 30, 58, 105, 108, 109, 110, 113, 114
Katz, 78
Kauffmann, 95
Kepler, 62
Khayyām, 8, 15, 78
Killing, 5
Kircher, 11
Klein, 121
Klügel, 105, 106, 108, 109, 113
Knar, 113
Knecht, 4
Knobloch, vi, 64, 71, 82, 85
König, 103, 104, 118

L
Lagrange, 118, 119, 120
Lambert, v, 6, 17, 29, 39, 50, 57, 98, 104, 105, 106, 108, 111, 117, 118
Lauter, 5
Le Blonde, 84
Lechalas, 60
Legendre, 57, 74, 114
Lehmann, 117
Levey, 68
Lévy, 10, 75
L'Hospital, 64, 84
Lobachevsky, 52, 60, 87, 106, 109, 120
Lo Bello, 8, 9, 93
Locke, 35
Loemker, 33, 36, 37
Look, 49
Lorenz, 111

M
Maierù, 18
Maimon, 12
Majer, 5
Malebranche, 23, 24, 35, 67
Mancosu, 94, 113
Mantovani, vi
Marchetti, 15, 16, 64
Mariotte, 54, 118
Maupertuis, 104, 118
McClenon, 78
Medici, 9
Meissner, 4
Mellizo, 81
Mercator, 43, 76, 95, 109
Merry, vi
Mesnard, 70
Miller, 92
Monge, 82
Mueller, 84
Mugnai, vi, 46, 82, 84
Münzenmayer, 4
Mylon, 19

N
Napolitani, 9, 81
Nasīr ad-Dīn at-Tūsī, 8, 9, 10, 11, 13, 16, 17, 23, 64, 92, 111
Newton, 42, 43, 44, 51, 85
Nicole, 23
Nizolius, 34, 65

O
Ohm, 116, 117
Olaso, 81

Oldenburg, 13
Oliveira, 119

P
Pambuccian, vi, 92
Pascal, 23, 30, 32, 62, 64, 69, 70, 129
Pasch, 63, 119
Pasini, vi
Patrizi, 42, 43, 63, 68
Peano, 72
Peletier, 12, 37
Pelletier, 108
Pepe, 119, 120
Perier, 69
Piro, vi
Playfair, 68, 89, 119
Pocock, 16, 17
Pont, 7, 117, 118
Posidonius, 9, 29
Probst, vi, 19, 76, 80
Proclus, 5, 7, 8, 9, 10, 12, 16, 18, 24, 25, 26, 29, 36, 59, 60, 68, 75, 80, 81, 83, 92, 93, 94, 95, 96, 97, 101, 112, 137, 159, 167, 169, 175, 177
Ptolemy, 101
Pulte, 104
Pythagoras, 22, 83, 84

R
Rabouin, vi, 32, 78
Raimondi, 9, 82
Ramus, 12, 25, 80
Rashed, 14
Régis, 23
Reichardt, 116
Remnant, 38, 165
Rescher, 49
Richard, 63, 101
Riemann, 52, 121
Risse, 28, 79
Roberval, 13, 14, 25, 26, 60, 63, 95
Robinet, 42, 43, 51, 53, 54
Robinson, 64
Rosenfeld, 8
Russell, 46
Rutherford, 49

S
Sabra, 8
Saccheri, v, 6, 11, 12, 14, 15, 17, 18, 22, 31, 39, 50, 57, 60, 64, 66, 67, 69, 73, 78, 80, 83, 92, 95, 98, 100, 101, 105, 106, 108, 109, 110, 118, 119
Sánchez, 81, 155
Sauveur, 84, 85, 109
Savile, 4, 16, 19, 23, 24
Schenk, 59, 84
Schott, 11, 12, 21, 30, 71, 76
Schotten, 113
Schubring, 115
Schultz, 113, 114, 115, 116
Schwab, 111, 112, 116
Schweikart, 116
Segner, 101, 111
Sengmueller, vi
Sherry, 78
Simplicius, 8, 9, 93
Stäckel, 116
Staudt, 110
Stegmann, 53
Stirling, 93
Suarez, 81
Sude, 60

T
Tacquet, 12
Taton, 65, 120
Taurinus, 117
Tenca, 17
Thābit ibn Qurra, 8, 10, 14, 37
Thales, 26, 66, 77, 119
Tilling, 118
Tonelli, 107
Tschirnhaus, 35
Tummers, 9, 93

U
Unguru, 82

V
Vagetius, 24, 25, 28, 29, 33, 79
Varignon, 81, 120
Vermehren, 112
Veronese, 5, 44
Vitale Giordano, 15, 80, 93
Voigt, 109
Volkert, 113

W
Wallis, 11, 16, 17, 18, 19, 29, 50, 51, 62, 64, 73, 76, 105, 117

Wallwitz, 4
Webb, 108
Wolff, 29, 39, 77, 103, 105, 106, 107, 108, 112

Z
Zormbala, 60, 87

Kolibri 360 GmbH
Ferdinand-Jühlke-Straße 7
99095 Erfurt, Deutschland
produktsicherheit@kolibri360.de